MODELING AND SIMULATION
FOR COLLECTIVE DYNAMICS

LECTURE NOTES SERIES
Institute for Mathematical Sciences, National University of Singapore

Series Editors: Chi Tat Chong and Alexandre Thiery
Institute for Mathematical Sciences
National University of Singapore

ISSN: 1793-0758

Published

For the complete list of titles in this series, please go to
https://www.worldscientific.com/series/LNIMSNUS

Lecture Notes Series, Institute for Mathematical Sciences, National University of Singapore

Vol. 40

MODELING AND SIMULATION FOR COLLECTIVE DYNAMICS

Editors

Weizhu Bao

National University of Singapore, Singapore

Peter A Markowich

King Abdullah University of Science and Technology, Saudi Arabia

Benoit Perthame

Sorbonne Université, France

Eitan Tadmor

University of Maryland, USA

World Scientific

EW JERSEY · LONDON · SINGAPORE · BEIJING · SHANGHAI · HONG KONG · TAIPEI · CHENNAI · TOKYO

Published by

World Scientific Publishing Co. Pte. Ltd.

5 Toh Tuck Link, Singapore 596224

USA office: 27 Warren Street, Suite 401-402, Hackensack, NJ 07601

UK office: 57 Shelton Street, Covent Garden, London WC2H 9HE

Library of Congress Cataloging-in-Publication Data

Names: Bao, Weizhu, editor. | Markowich, Peter A., 1956– editor. |
 Perthame, B., editor. | Tadmor, Eitan, editor.
Title: Modeling and simulation for collective dynamics / editors: Weizhu Bao,
 Peter A. Markowich, Benoit Perthame, Eitan Tadmor.
Description: New Jersey : World Scientific, [2023] | Series: Lecture notes series,
 Institute for Mathematical Sciences, National University of Singapore 1793-0758 ; Vol. 40 |
 Includes bibliographical references.
Identifiers: LCCN 2022045047 | ISBN 9789811266133 (hardcover) | ISBN 9789811266140 (ebook)
Subjects: LCSH: Dynamics--Mathematical models. | System analysis. | Mathematical optimization.
Classification: LCC TA352 .M63 2023 | DDC 620.1/04011--dc23/eng/20221129
LC record available at https://lccn.loc.gov/2022045047

British Library Cataloguing-in-Publication Data
A catalogue record for this book is available from the British Library.

Cover image by Sophie Hecht.

For any available supplementary material, please visit
https://www.worldscientific.com/worldscibooks/10.1142/13136#t=suppl

Contents

Foreword

The Institute for Mathematical Sciences (IMS) organizes thematic programs of longer duration as well as shorter workshops and public lectures. The themes are selected from among areas at the forefront of current research in the mathematical sciences and their applications. Each volume of the *IMS Lecture Notes Series* is a compendium of papers based on lectures and tutorials delivered at the IMS. The aim is to make original papers and expository articles on a subject of current interest available to the international research community. These volumes also serve as a record of activities that took place at the IMS. We hope that through regular publication of these Lecture Notes the Institute will achieve, in part, its objective of reaching out to the community of scholars in the promotion of research in the mathematical sciences.

Chi Tat Chong
Alexandre Thiery
Series Editors

Preface

Quantum and kinetic models have been widely used in the modeling and description for many problems arising in science and engineering with quantum effect (wave-particle duality and/or quantization) and/or particle interaction. Over the last two decades, quantum and kinetic models have been adapted for the kinetic description of emerging applications in biology and social science, such as cell migration, collective motion of active matter, network formation and dynamics in social sciences, coherent structures in crowd and traffic dynamics, flocking, swarming, and for the modeling of tremendous new experiments in physics, such as Bose-Einstein condensation, fermion condensation, quantum fluids of light, degenerate quantum gas, graphene and 2D materials, etc. These new surprising experiments and emerging applications generate a big wave in the study of challenging *quantum and kinetic problems* in terms of modeling, analysis and simulation. In fact, the new experiments and applications also call for greater participation of mathematicians and computational scientists to address some fundamental questions related to quantum and kinetic problems, to work together with applied scientists from the modeling to computational stages, to provide mathematical analysis for justifying different models, and to design efficient and accurate computational methods. A thematic program on *Quantum and Kinetic Problems: Modeling, Analysis, Numerics and Applications* was held at the Institute for Mathematical Sciences (IMS) at the National University of Singapore (NUS) from September 2019 to March 2020. The principle goals of this thematic program were to:

- present recent developments in mathematical theories, including modeling, analysis and computational techniques that is relevant to quantum and kinetic problems;
- discuss and compare different, recent-proposed quantum and kinetic models related to the latest emerging applications;
- identify critical scientific issues in the understanding of quantum

and kinetic problems and difficulties of common interest within two
different disciplines as well as issues that are specific to individual
area;

- accelerate the interaction of mathematicians and applied scientists
  by stimulating lively debate on important research issues related
  to quantum and kinetic problems, and promote highly interdisci-
  plinary research with emerging applications and cross fertilization
  of ideas;
- develop and foster international and local collaborations of scientific researches in quantum and kinetic problems;
- organize focused panel discussions that assemble panelists from
  a diverse research spectrum to lead further dialogue between researchers with different training and expertise; and
- help to train junior researchers and graduate students by exposing
  them to a broad area of mathematical knowledge and computational techniques through tutorial lectures, public lectures, research
  seminars and collaborations.

The thematic program included a period of collaborative research, five
week-long workshops, several distinguished lectures, and a number of tutorials. As an important part of the program, tutorials and special lectures were
given by leading experts in the fields for participating graduate students
and junior researchers. They covered a number of aspects on modelling
and simulation for collective dynamics including individual and population approaches for population dynamics in mathematical biology, collective behaviors for Lohe type first-order aggregation models, mean-field
particle swarm optimization, and consensus-based optimization and ensemble Kalman inversion for global optimization problems with constraints.
The current volume *Modeling and Simulation for Collective Dynamics* col-
lects four expanded lecture notes with each self-contained tutorials. The
following is a brief introduction to these lecture notes:

Chapter 1. Individual and Population Approaches for Calibrating Division
          Rates in Population Dynamics: Application to the Bacterial
          Cell Cycle by Marie Doumic and Marc Hoffmann: It reviews
          the main results developed over the last decade for modeling,
          analysing and inferring triggering mechanisms in population
          reproduction which is an active and growing research domain
          in mathematical biology. It focuses on different methods for
          the estimation of the division rate in growing and dividing

populations in a steady environment. These methods combine tools borrowed from PDE's and stochastic processes, with a certain view that emerges from mathematical statistics. It provides a concrete presentation on the application to the bacterial cell division cycle and helps the reader to identify major new challenges in the field.

Chapter 2. Collective Dynamics of Lohe Type Aggregation Models by Seung-Yeal Ha and Dohyun Kim: It reviews state-of-the-art results on the collective behaviors for Lohe type first-order aggregation models. It focuses on two Lohe type models: the Lohe tensor model and the Schrödinger-Lohe (SL) model, and presents several sufficient conditions in unified frameworks via the Lyapunov functional approach for state diameters and dynamical systems theory approach for two-point correlation functions. Several numerical simulation results are reported for the SL model.

Chapter 3. Mean-Field Particle Swarm Optimization by Sara Grassi, Hui Huang, Lorenzo Pareschi and Jinniao Qiu: It surveys some recent results on the global minimization of a non-convex and possibly non-smooth high dimensional objective function by means of particle based gradient-free methods, arising in machine learning and signal processing. Based on particle swarm optimization (PSO), it introduces a continuous formulation via second-order systems of stochastic differential equations, shows to derive in the limit of large particles number via Vlasov-Fokker-Planck type equations, and analyzes the corresponding macroscopic hydrodynamic equations in the zero inertia limit. Rigorous results concerning the mean-field limit, the zero-inertia limit, and the convergence of the mean-field PSO method towards the global minimum are provided along with a suite of numerical examples.

Chapter 4. Consensus-based Optimization and Ensemble Kalman Inversion for Global Optimization Problems with Constraints by Jose Antonio Carrillo, Claudia Totzeck and Urbain Vaes: It introduces a practical method for incorporating equality and inequality constraints in global optimization methods based on stochastic interacting particle systems, specifically consensus-based optimization and ensemble Kalman inversion. The properties of the method are studied through the associated mean-field Fokker–Planck equation and its performance is demonstrated by numerical experiments on several test problems.

There were many colleagues who had helped putting the program together. Besides four of us as co-chairs of the Organizing Committee of this thematic program, other members of the organizing committee include: José Antonio Carrillo (University of Oxford), Ionut Danaila (Université de Rouen Normandie), Yuan Ping Feng (National University of Singapore), Dieter Jaksch (University of Oxford), Shi Jin (Shanghai Jiao Tong University), Henrik Jönsson (University of Cambridge), Choy Heng Lai (National University of Singapore), Mark Lewis (University of Alberta), Christian Lubich (Universität Tübingen), Antonio Helio Castro Neto (National University of Singapore), Lorenzo Pareschi (University of Ferrara), and Zhouping Xin (The Chinese University of Hong Kong). We are very much grateful to their invaluable services. Thanks also to all the participants of this program for their support and stimulating interactions during the six months!

We would like to take this opportunity to thank Professor Chi Tat Chong, Director of IMS, for his leadership in creating an exciting environment for mathematical research in IMS and for his guidance throughout our program. The expertise and dedication of all IMS staff contributed essentially to the success of this program. Last but not least, we would like to acknowledge IMS for providing financial support to the program as well as the lecture room, the office and work space in the enthusiastic environment of NUS.

November 2022

Weizhu Bao
National University of Singapore

Peter A. Markowich
King Abdullah University of Science and Technology

Benoit Perthame
Sorbonne Université

Eitan Tadmor
University of Maryland

Volume Editors

https://doi.org/10.1142/9789811266140_0001

Individual and Population Approaches for Calibrating Division Rates in Population Dynamics: Application to the Bacterial Cell Cycle

Marie Doumic* and Marc Hoffmann[†]

*Sorbonne Universités, Inria, UPMC Univ Paris 06
Lab. J.L. Lions UMR CNRS 7598, Paris, France
marie.doumic@inria.fr

[†]University Paris-Dauphine, CEREMADE,
Place du Maréchal De Lattre de Tassigny, 75016 Paris, France
hoffmann@ceremade.dauphine.fr

Modelling, analysing and inferring triggering mechanisms in population reproduction is fundamental in many biological applications. It is also an active and growing research domain in mathematical biology. In this chapter, we review the main results developed over the last decade for the estimation of the division rate in growing and dividing populations in a steady environment. These methods combine tools borrowed from PDE's and stochastic processes, with a certain view that emerges from mathematical statistics. A focus on the application to the bacterial cell division cycle provides a concrete presentation, and may help the reader to identify major new challenges in the field.

Keywords: cell division cycle, bacterial growth, inverse problem, nonparametric statistical inference, kernel density estimation, growth-fragmentation equation, growth-fragmentation process, renewal equation, renewal process, adder model, incremental model, asymptotic behaviour, long-term dynamics, eigenvalue problem, Malthusian parameter

Mathematics Subject Classification: 35R30, 92B05, 35Q62, 62G05

Contents

1. Introduction

1.1. *Biological motivation*

The study of stochastic or deterministic population dynamics, their qualitative behaviour and the inference of their characteristics is an increasingly important research field, which gathers various mathematical approaches as well as application fields. It benefits from the huge advances in gathering data, so that it is today possible not only to write and study qualitative models but also to calibrate them and assess their relevance in a quantitative manner. This chapter aims at contributing to review some recent advances and remaining challenges in the field, through the lens of a specific application, namely the bacterial cell division cycle. Guided by this application, we propose here a kind of roadmap for the mathematician in order to tackle a genuinely applied problem, coming from contemporary biology.

How does a population grow?

Let us begin by describing the growth of a microcolony of bacteria, illustrated in the snapshots of Fig. 1: out of one rod-shaped *E. coli* bacterium, a colony rapidly emerges by the growth of each bacterium and its splitting into two daughter cells.

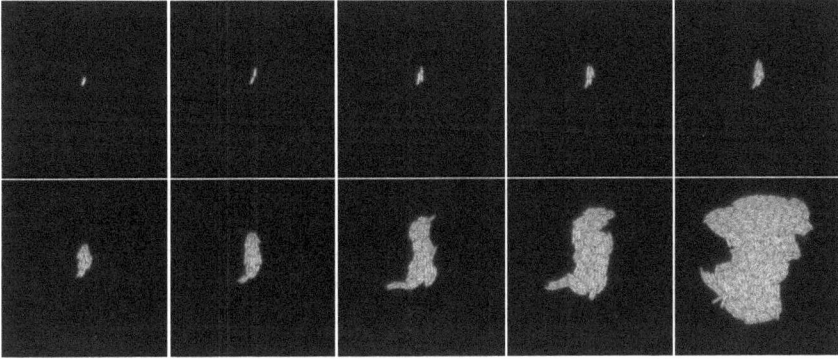

Fig. 1.   From left to right and top to bottom: Successive snapshots of the video https://doi.org/10.1371/journal.pbio.0030045.sv001.

Nutrient being in large excess at this development stage, we can assume a steady environment. We also ignore the many fascinating questions arising from spatial considerations [1–3], and focus on the two fundamental mechanisms at stake: growth and division. We can go back and forth between the population and the individual view: how does the knowledge of the growth and division laws of the individual lead to the knowledge of the population growth law, and conversely, to which extent observations on the population can help us infer the individual laws?

How does a cell divide?

To follow more easily individual characteristics of the cells over many generations, a microfluidic liquid-culture device called "the mother machine" has been developed in the last decade with an increasing success [4,5]. This is illustrated in Fig. 2, a snapshot taken from the illustrating Movie S1, from Ref. [5]. The population is then reduced to independent lineages, but the two main mechanisms remain the same: growth and division. How do these two mechanisms coordinate each other? What triggers the bacterial

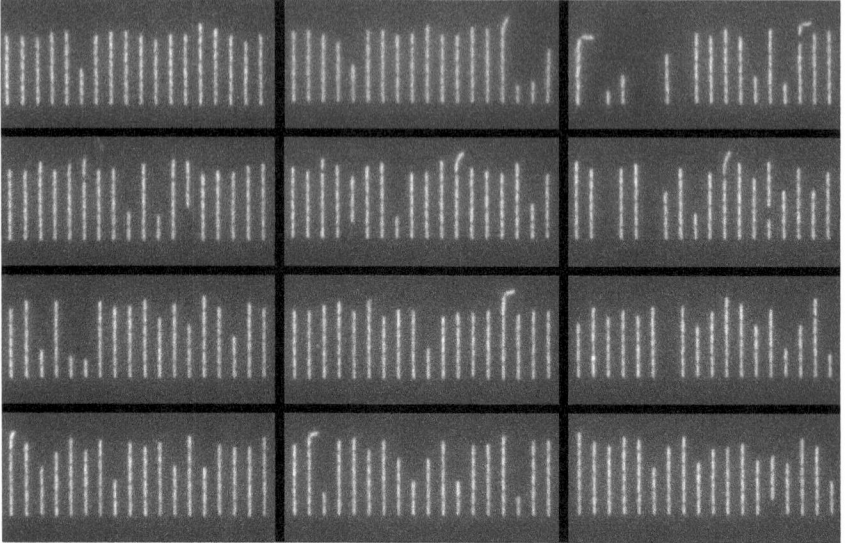

Fig. 2. From left to right and top to bottom: Snapshot of the video S1 of Ref. [5]. In each channel bacteria grow and divide while remaining aligned, pushing new born cells towards the exit.

division? To answer such questions, many studies have deciphered complex intracellular mechanisms, see for instance the recent review [6], while others aim at inferring laws of growth and division out of the observation of population (as in Fig. 1) or lineages (as in Fig. 2) dynamics. This last approach, which could be named phenomenological rather than mechanistic, constitutes the guideline of this chapter, and could be summed-up by the problematic: How much information on triggering mechanisms of growth and division can be extracted from such data as in Figs. 1 and 2?

Other applications

We follow here the application to cell division; however, many other are possible, such as polymer fragmentation [7–9], or mineral crushing [10], or yet other types of cell division cycles [11]. This would lead us too far for this chapter, but we believe that many ideas gathered here for bacteria may apply to other fields.

1.2. *Outline of the chapter*

To serve as an outline of the chapter, let us enumerate the main steps towards the formulation of laws of growth and division in a process which is rather circular than linear in practice. This — admittedly subjective — methodological guideline is quite general, and many readers should recognize their own approach to their own problem; we then explain and specify them when applied to growing and dividing populations, and, more precisely, to the bacterial cell division cycle.

- **Step 1) Preliminaries** — Secs. 1.3 and 1.4
  - **Analyse data:** (or make the most of direct observations)
    Biological data are often extremely rich, and only part of this richness is effectively analysed by experimenters, for instance through the use of averaged quantities rather than individual measurements. At the same time, the noise level is often very high, requiring an appropriate noise modelling approach. Mathematical methods at this first step are a combination of statistics, image analysis and interdisciplinary discussions between modellers and experimenters. This is carried out in our application case to bacterial growth and division in Sec. 1.3.
  - **Specify assumptions:** (or "model and simplify")
    The data analysis carried out at Step 1 should lead to two types of hypothesis: simplifying ones, *a priori* justified by statistical quantifiers and which should be *a posteriori* verified by sensitivity analysis; and modelling ones, guiding conjectures on the underlying laws, which should be challenged at the end of the procedure. This is carried out in Sec. 1.4.

- **Step 2) Build a model** — Sec. 2
  The goal of Sec. 2 is to translate mathematically the assumptions done at Step 1. As suggested by the illustrations of Figs. 1 and 2, we may distinguish two types of models: individual-based and population models, leading to stochastic processes, branching trees, or integro-partial differential equations (PDEs).

- **Step 3) Analyse the model** — Sec. 3
  The models analyses, carried out in Sec. 3, could seemingly be skipped to go directly to the conclusion by simulating the models as best as possible, with available simulation packages, and

comparing them to the data, using here again available fitting tools. We believe however that such approaches not only lack rigour but also risk missing enlightening information. Many success stories in many application domains could illustrate this general comment; in our very field, the cornerstone and foundation of the inverse problem solutions lies in the long-time asymptotics of the population models, as first described by B. Perthame and J. Zubelli [12]. For the polymer breakage application, not developed in this chapter, the use of time asymptotics as proved in Ref. [13] drastically simplified and justified the calibration of the fragmentation model [8,9]. In Sec. 3, we thus review some of the main theoretical results which may prove useful for the following steps.

- **Step 4) Calibrate the model** — Sec. 4
  We can do here the same remark as for Sec. 3: standard calibration methods are often applicable, reducing this step to the choice of up-to-date softwares. A main drawback would be the difficulty to assess the confidence we may have in the results obtained; above all, theoretical error estimation inform us on the quantity of information available in the data collected, thus able to inspire design of new experiments. We thus develop in Sec. 4 the inverse problem analyses carried out in the past decade for the different observation schemes, detailing some of the proofs in the illuminating example of the renewal model and in the "ideal mitosis" case, and reviewing the results obtained for more complex cases.

- **Step 5) Conclusion** — Sec. 5
  At this stage, we have all the necessary tools on hand to confront a model to the data, and conclude on the validity on the modelling assumptions made at Step 2. In Sec. 5, a protocol to confront model to data is proposed and applied to experimental data for bacteria. It is however a step where many questions remain open, concerning the design and analysis of statistical tests as well as the formulation of new or more detailed models. It usually lead to boostrapping the methodology by circulate another round from Steps 1 to 5.

1.3. *First step: data analysis*

To build models that can be compared to the data, a preliminary step consists in having clear ideas on what can be measured and how. We first distinguish two types of data collection and two observation schemes, and then give some examples of what can be directly obtained from the data.

Data description: two types of datasets

We have already shown in Figs. 1 and 2 two modern experimental settings to observe bacterial growth. In these two settings, a picture is taken at given time intervals — typically here, every 1 to 5 min — giving access, through image analysis, not only to time-dependent (noisy) samples of sizes but also to genealogical data and to the knowledge of the time elapsed since birth, of size dimensions at birth and of size dimensions at division, and of the ratio between the mother cell size at division and the offspring sizes. In the sequel, we call such cases *individual dynamics data*, meaning that some knowledge about the growth and division processes may be directly inferred from the data, where individual dynamics are collected.

However, there are other situations, for instance when *ex vivo* samples are collected, when only cheaper devices are available, or when the quantities of interest are not dynamically measured, or yet for other applications such as protein fibrils. In all these cases, the experimenter can only observe size distribution of particles of interest, taken at one or several time points, but without being able to follow each one so that no individual observation of growth or division may be done. In the sequel, we call such cases *population point data*, meaning that we have information on the population dynamics or on point individual data, but no access to individual dynamics.

As detailed in Sec. 4, each type of data collection raises different problems. Schematically, in the population point data, one has to choose a given model, and the estimation questions at stake are to determine which parameters may be inferred from the data and how precise this inference is. In the individual dynamics data, data are much richer and so are the questions to tackle: first, as for the population point data, estimate model parameters, and second, assess quantitatively how accurate the model is, evaluate to which extent it can be enriched without over-estimation, and compare it with other models.

Data description: genealogical observation vs. population observation

In the individual dynamics data cases, we have seen two distinct experimental settings: either we follow the overall population until a certain time, as in Fig. 1 — we call this case the *population case*, corresponding to $k = 2$ in the following, k being the number of children at division — or we follow only one given lineage, since at each division we keep observing one out of the two daughter cells — we call this case the *genealogical case*, corresponding to $k = 1$ in the following. As explained in Sec. 2, the mathematical model needs to adapt to these two cases, and so need the mathematical analysis and the model calibration.

At first sight, the population point data collection seems to apply only to the population observation scheme ($k = 2$), taking sparse pictures of a population state at some times. However, it could be imagined that in a given genealogical observation experiment ($k = 1$) we are not able to determine when or where a cell divides; this will be the case for instance if the timesteps of observation are too large. It is thus also interesting to develop models and calibration methods for this seemingly strange but not unrealistic case.

For the population observation case $k = 2$, if the cells are in constant growth conditions (unlimited nutrient and space), the so-called Malthusian parameter characterises the exponential growth of the population. We denote it here λ, meaning that the population grows like $e^{\lambda t}$ — rigorous meanings of this statement are provided in Sec. 3. The Malthusian parameter can be measured in various ways for many different experimental conditions — as the total biomass increase for instance. Equivalently, biologists often refer to the doubling time T_2 of the population, with the immediate relation $T_2 = \frac{\ln(2)}{\lambda}$.

Data analysis: size distributions

To each observation scheme correspond different types of data and specific measurement noise. Let us here gather some of the most frequent information we can extract.

It is possible to extract size dimension distributions from the two types of data collection described above. For instance, for a given sample of n cells in a *E. coli* population, we measure their lengths at a given time t to obtain

$$\big(x_1(t), \cdots, x_n(t)\big),$$

the width being considered roughly constant among cells. As a first approximation, we may assume that the observation is a n-drawn of a random vector

$$\big(X_1(t), \cdots, X_n(t)\big),$$

where the random variables $X_i(t)$ are independent, with common distribution $\mu(t)$; this assumption is obviously not valid, since we ignore the underlying dependent structure. Yet, it may well happen — and this will be extensively discussed later — that the sample behaves approximately for certain linear statistics like an n-drawn of a common distribution [14]. In particular, we may at least expect the convergence of the empirical measure

$$\mu_n = \frac{1}{n} \sum_{k=1}^{n} \delta_{X_k(t)} \tag{1.1}$$

to $\mu(t)$ in a weak sense [15], possibly quantified by a distance like Wasserstein [16,17] that metrizes weak convergence. Assuming that μ has a density, and using for instance histograms or adaptive kernel density estimators we may obtain a smooth estimate for the size distribution $\mu(t)$. This is treated at length in Sec. 4. Assuming moreover that $\mu(t) = \mu$ does not depend on time, an assumption that can (and will) be justified in some cases by the asymptotic analysis carried out in Sec. 3, also known in biology as cell size homeostasis, we may concatenate all data taken at all time to get larger samples. This is illustrated in Fig. 3(Right), where we have shown the length-distribution of cells taken at any time out of genealogical observation ($k = 1$, data from Ref. [5]) in blue and population observation in green ($k = 2$, data from Ref. [18]).

Data analysis: individual dynamics data analysis

In the case of individual dynamics data collection, it is not only possible to extract size distribution but also much more. Let us list some of the information we can extract from such rich data.

To begin with, we can measure the cell age, *i.e.* the time elapsed since birth; or yet — let us mention it due to its recent importance in the field [20–22] — size increment, *i.e.* the difference between the size of the bacteria at the time considered and their size at birth. A similar process as seen above for size leads us to age or size increment distributions as illustrated in Fig. 3(Left). In the same vein, we can also select only dividing cells, measure their size, age, size increment at division, and estimate these distributions:

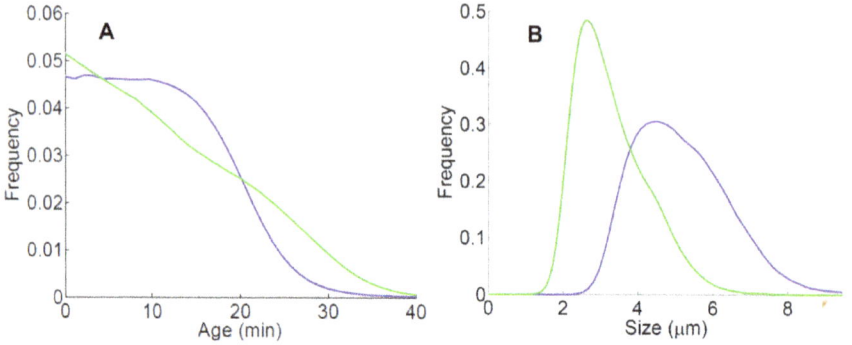

Fig. 3. **All cells distributions:** Kernel density estimation of age (Left) and length (Right) distributions obtained from sample images of genealogical observation (blue curves, data from Ref. [5]) and population observation (green curves, data from Ref. [18]), data taken at all time points. Figure taken from Ref. [19], Fig. 1.

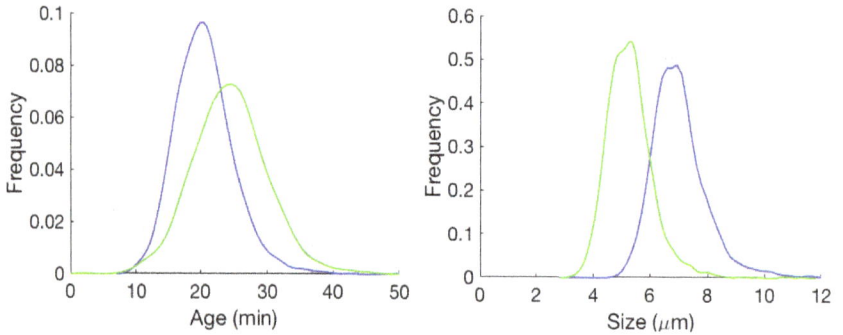

Fig. 4. Dividing cells distributions: Kernel density estimation of age (Left) and length (Right) distributions of dividing cells obtained from sample images of genealogical observation (blue curves, data from Ref. [5]) and population observation (green curves, data from Ref. [18]).

this is illustrated in Fig. 4. Finally, we can measure joint distributions, such as age-size or size-increment of size distributions: this is done in Fig. 5.

Being able to follow each cell means that we are also able to estimate individual growth rates. This is illustrated for one given cell in Fig. 6: measuring cell length and cell width every two minutes allows one to estimate its growth rate by curve fitting tools. On the right, we see width measurements: it appears to remain constant up to measurement noise. On the left, we see the fit of the data with an exponential curve in red and with a

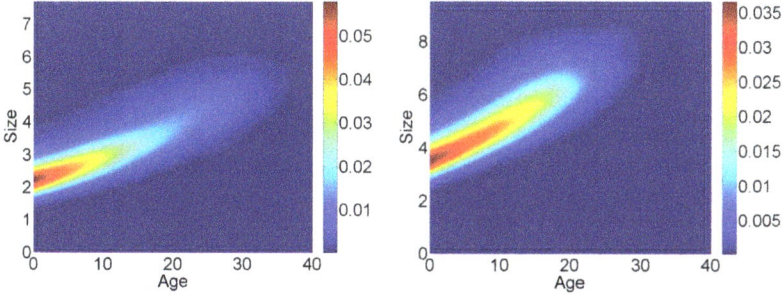

Fig. 5. Age-Size distributions for all cells. Left: population observation, data from Ref. [18], Right: genealogical observation, data from Ref. [5].

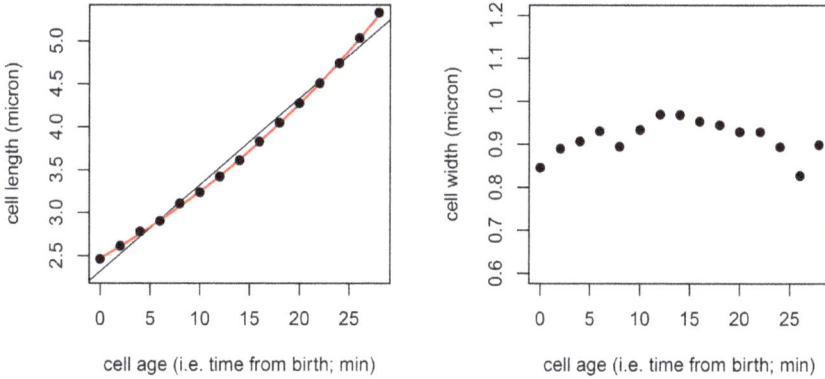

Fig. 6. Single-cell growth rate analysis. Figure taken from Ref. [19] Fig. 2. For a given cell, we measure its size dimensions — length on the right, width on the left — through time, and conclude to a good agreement of the exponential growth model $\frac{dx}{dt} = \kappa x$ for the length, and a constant width.

linear curve in black: as studied in Ref. [19], and in accordance with previous studies, the exponential growth model, where we assume that the cell length evolves according to the law

$$\frac{dx}{dt} = \kappa x$$

for a certain rate $\kappa > 0$, fits the data very well — at least for the lengths where we have data, *i.e.* here for a range of sizes, typical size between 0.3 to $5\mu m$.

But is this growth rate constant among all cells? As for the age or size distributions, once estimated for each cell, and assuming them independent (no heritability) and time-independent (no slowing down in growth due to lack of nutrient for instance), it is possible to study the distribution of growth rates. This is illustrated in Fig. 7(Left).

Concerning division, we already mentioned that it is possible to extract dividing cells distribution. But we have more information: for instance, it is possible to measure the ratio between daughter and mother cells. In our application setting, this is called the septum position, the *septum* being the boundary formed between the two dividing cells.

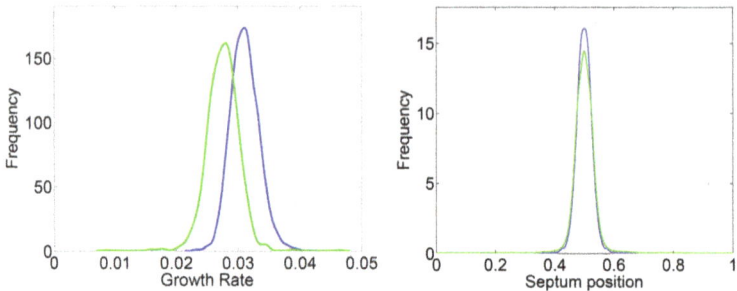

Fig. 7. Growth rate in $\min^{-1}$ (Left) and daughter/mother ratio (Right) distributions. Blue curves: genealogical data from Ref. [5], Green curves: population data from Ref. [18]. Figure taken from Fig. S4 of Ref. [19].

1.4. *Second step: making assumptions*

In the first step we have started to manipulate the data. It is clear that we could still get a lot of information from them: inheritance between mother and daughter cells, distributions over time and not just aggregating all the time data, etc. Here and there, we already made two types of assumptions: model assumptions and simplifying assumptions. Let us gather them here, so that we will be ready to design mathematical models.

Simplifying assumptions can be made out of the direct observations done during the first step. In our application case, we list the following assumptions which will be used as a departure point for the calibration step, see Sec. 4.

- The daughter cell size at birth is half its mother cell size (Fig. 7(Right) shows very little variability: we may neglect it first).

- All cells grow exponentially with the same growth rate κ (Fig. 7(Left) shows some variability, that can be neglected in a first approximation).
- Space and nutrient consumption are infinite and do not influence growth and division (due to the experimental setting, this assumption may be verified by statistical analysis of time dynamics).

Model assumptions of course depend on the underlying application. They are the gateway to the third step and a way to formulate the vague question of the introduction: how to determine laws for growth and division? In our case, a central assumption, linked to the Markov property of our model, is the absence of memory between mother and daughter cells, *i.e.* we assume no heritability of the growth rate — see Ref. [23] for a thorough study of this question — and no heritability of the division features. Concerning growth, in the case of individual dynamics data collection, we have seen that we can formulate laws directly inferred from the data. This is not true concerning the law of division: the question "what triggers bacterial division?" remains unanswered by our data analysis step. We thus formulate the following modelling assumptions. They provide guidelines for all our subsequent models:

- A particle of age a and size x may divide with a division rate B depending on its age,
- a particle of age a and size x may divide with a division rate B depending on its size,
- a particle of size x and size at birth x_b may divide with a division rate B depending on its increment of size $x - x_b$,
- a particle of size x may divide with a division rate B depending on an auxiliary (and latent, *i.e.* unobserved) variable.

2. Building models

The preliminary steps sketched in Secs. 1.3 and 1.4 allow one to have a clear idea on the necessary ingredients to translate mathematically the biological mechanisms to study. We list three apparently different approaches:

1. **Continuous-time branching processes:** the most direct and intuitive way is to model each cell of the population inside its genealogical tree, linking the parent to its offspring by a tree branch representing its lifetime, each node representing a cell taken at birth or at division. We explain this model below in Sec. 2.1.

2. **Stochastic differential equations (SDE) via Poisson random measures:** this formalism is strictly equivalent to the building of the branching tree and consists in writing a stochastic differential equation satisfied by a random measure representing all the cells alive at a certain time. The advantage of writing this equation is that it is a very convenient way to link the stochastic model to the "deterministic" — or, more adequately, average — approach described in Sec. 2.3. This limit is rigorously proved in Sec. 2.2 in the pedagogical case of the renewal process, and we review results of the literature for other models.

3. **Integro-partial differential equations (PDE):** looking at a *large* population, or yet at the *average* behaviour of one or of a small number of individuals, we can write a balance equation satisfied by the concentration distribution of cells at time t, with given characteristics such as age, size, increment of size, etc.: this gives rise to what is called a *structured population equation*, the term *structured* referring to this characteristic *trait* triggering growth, division, or more generally speaking evolution/change. This type of models is reviewed in Sec. 2.3.

Depending on the context or on the specific questions to be solved, one of the three above points of view may seem more advantageous, either from a technical or an interpretation point of view. The three approaches are closely related and sometimes equivalent. We next describe somehow their minimal mathematical features and their links.

2.1. *Continuous-time branching processes*

Continuous-time branching processes are classical objects, well documented in numerous textbooks and papers, see *e.g.* Refs. [24–27] or Refs. [28–31]. Reference [32] is recommended for an efficient presentation of the topic.

The models we present here belong to the wide family of so-called "continuous-time branching trees", whose history dates back to 1873 [25] and knows continuous interest from ecologists as well as mathematicians, see *e.g.* Ref. [28] or for a specific and very recent example [33]. For the sake of simplicity, we stick here to the modelling and simplifying assumptions sketched above, so that our branching process is encoded in a binary tree: each node splits into exactly two branches. Using the Ulam-Neveu notation, we define

$$\mathcal{U} := \bigcup_{n=0}^{\infty} \{0,1\}^n \quad \text{with} \quad \{0,1\}^0 := \emptyset. \tag{2.1}$$

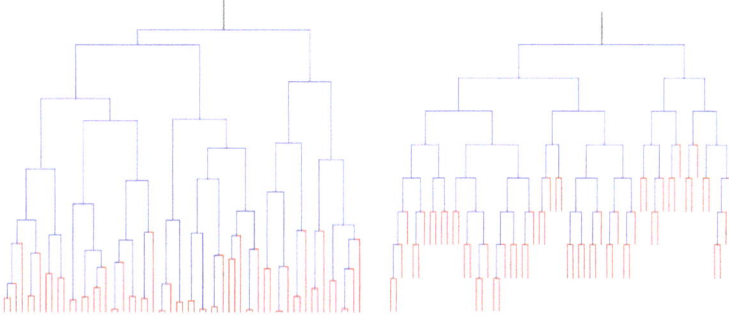

Fig. 8. Simulation of a binary tree with. Left: the size of each segment represents the lifetime of an individual. Individuals alive at time t are represented in red. Right: genealogical representation of the same realisation of the tree. Figure taken from Fig. 1 of Ref. [14].

To each node $u \in \mathcal{U}$, we associate a cell with birth time b_u, size at birth ξ_u, lifetime ζ_u and increment of size since birth η_u. Assuming that the size growth is given by the growth rate $\tau(x)$ (in the sequel, we will often specify $\tau(x) = \kappa x$ for some common growth rate $\kappa > 0$), if $u^- \in \mathcal{U}$ denotes the parent of $u \in \mathcal{U}$, then the size at division is given by $\chi_u = X(\zeta_u, \xi_u)$ where $X(t, x)$ denotes the characteristic curve solution to the differential equation

$$\frac{dX(t, x)}{dt} = \tau\left(X(t, x)\right), \qquad X(0, x) = x,$$

and we have, for a division probability kernel $b(\chi_u, dx)$

$$\xi_u \sim b(\chi_{u^-}, dx), \qquad \eta_u = \chi_u - \xi_u = X(\zeta_u, \xi_u) - \xi_u.$$

In the specific case of exponential growth and diagonal kernel (equal mitosis, for which $b(\chi_u, dx) = \delta_{\frac{\chi_u}{2}}(dx)$), this gives

$$\xi_u = \frac{\chi_{u^-}}{2} = \frac{\xi_{u^-}}{2} \exp\left(\kappa\zeta_{u^-}\right), \qquad \eta_u = \xi_u \left(\exp(\kappa\zeta_u) - 1\right).$$

We see that within such a formalism, it is straightforward to generalise our assumptions: for instance, the growth rate κ could be selected randomly at birth, according to some probability law which could depend on the parent growth rate κ_{u^-} (or on some other trait of the mother). Another way to model unequal division is to write

$$\xi_{(u^-, 0)} = \alpha\xi_{u^-} \exp(\kappa\zeta_{u^-}), \qquad \xi_{(u^-, 1)} = (1 - \alpha)\xi_{u^-} \exp(\kappa\zeta_{u^-}),$$

with $\alpha \in (0,1)$ chosen randomly according to some probability distribution $b_0(d\alpha)$, symetric in $\frac{1}{2}$. In such a case, the probability kernel b has a specific form, which is sometimes called *self-similar*, namely

$$b(y,x) = \frac{1}{y}b_0\left(\frac{x}{y}\right).$$

One last ingredient is still missing to have fully determined the tree: the distribution of the random time at which division occurs. Let us list the three examples discussed in the assumptions section above:

1. **The division depends on age:** this translates into the fact that given a division rate function

$$B : (0,\infty) \to [0,\infty), \quad \int^\infty B(s)ds = \infty,$$

   the lifetime ζ_u is a random variable with distribution

$$\mathbb{P}(\zeta_u \in da) = B(a)e^{-\int_0^a B(s)ds}da.$$

   Other said, we have

$$\mathbb{P}(\zeta_u \in (a,a+da)|\zeta_u \geq a) = B(a)da, \quad \mathbb{P}(\zeta_u \geq a) = \exp\left(-\int_0^a B(s)ds\right).$$

   With this construction, the renewal process is simply embedded into a branching tree representation. See Ref. [14] for a study of a slight generalisation of this model. We notice that we can add other variables as the size and the increment of size, but they will have no influence on the tree.

2. **The division depends on size:** when the division is triggered by a size-dependent rate, and when the size grows at a rate $\tau(x)$, the division rate function now translates into the following properties

$$\mathbb{P}(\chi_u \in (x,x+dx)|\chi_u \geq x) = B(x)dx = \tau(x)B(x)dt,$$

$$\mathbb{P}(\chi_u \geq x|\xi_u) = \mathbb{1}_{\{x\geq\xi_u\}}\exp\left(-\int_{\xi_u}^x B(y)dy\right).$$

   Notice that the age variable ζ_u, although well-defined, is a bit irrelevant in this context: this is due to the fact that we define an instantaneous rate $B(x)dx$ instead of defining a rate $B(x)dt$. Compared to previous mathematical papers [34–37] etc., we simply substitute B by $B\tau$.

3. **The division depends on the increment of size since birth:** as for the renewal process, the increment is reset to zero at each division,

however if the growth rate $\tau(x)$ is not constant it does not increase linearly with time, so that η_u is now characterised by

$$\mathbb{P}(\eta_u \in (z, z+dz)|\eta_u \geq z, \xi_u) = B(z)dz = B(z)\tau(\xi_u + z)dt,$$

$$\mathbb{P}(\eta_u \geq 1) = \exp\left(-\int_0^z B(y)dy\right).$$

Indeed, for a size increment z we have a size since birth equal to $x = \xi_u + z$, and the size grows according to $\frac{dx}{dt} = d(\xi_u + z) = dz = \tau(x)dt$. When we embed the model in the time-continuous framework, we see that the lifetime does depend on size, contrarily to the renewal process, due to the fact that $dz \neq dt$ formally.

Given observational data, there are basically two ways of considering the tree: either we look at the process of structured cells *until a certain generation n* — along one branch chosen at random at each node (genealogical observation), or along chosen branches, or yet along all of them; alternatively, we look at the process *until a certain physical time* (population observation). In the first point of view, the physical time is not intrinsically important, contrarily to the second point of view, that we next adopt to describe the process via random measures.

2.2. *Random measures*

We consider the random process

$$X(t) = \big(X_1(t), X_2(t), \dots\big)$$

that describes the (ordered) sizes of the population at time t, or

$$A(t) = \big(A_1(t), A_2(t), \dots\big)$$

the (ordered) ages of the population at time t. Equivalently, we can define the random processes with values in finite-point measures on $(0, \infty)$ via

$$Z_t^{(a)} = \sum_{i=1}^\infty \delta_{A_i(t)}, \quad Z_t^{(x)} = \sum_{i=1}^\infty \delta_{X_i(t)}, \tag{2.2}$$

(the sum is finite but the total number of particles may be different for different values of t), that describe the population states, characterised by the structuring variables (here size, age, size increment and so on) at any given time t.

We can then look for a characterisation of the process $(Z_t^{(a)})_{t\geq 0}$ or $(Z_t^{(x)})_{t\geq 0}$ as via a stochastic evolution equation, here a measure-valued stochastic differential equation (SDE). In order to do so, we need a technical tool, namely the use of Poisson random measures.

Preliminaries: Poisson random measures

A convenient way to model scattered points or events through time and a state space is by means of a Poisson random measure. The notion and its properties are well-documented in numerous textbooks, from stochastic geometry to stochastic calculus, see *e.g.* Refs. [38–40] and we briefly recall the essential and basic material needed here.

We start with a state space $\mathcal{X} \subset \mathbb{R}^d$ and we let μ be some sigma-finite measure on $\mathcal{X}$ equipped with its Borel sigma-field. If $(x_i)_{i \in \mathcal{I}}$ is a countable family of elements of $\mathcal{X}$, the point measure N associated with $(x_i)_{i \in \mathcal{I}}$ is given by

$$N(dx) = \sum_{i \in \mathcal{I}} \delta_{x_i}(dx),$$

and its action on test functions $\varphi : \mathcal{X} \to \mathbb{R}$ is written as

$$\langle N, \varphi \rangle = \int_{\mathcal{X}} \varphi(x) N(dx) = \sum_{i \in \mathcal{I}} \varphi(x_i)$$

whenever the above sum is well-defined.

Definition 2.1: A random point measure N is a random Poisson measure on $\mathcal{X}$ with intensity μ if

1. For every Borel set A, the random variable $N(A)$ has a Poisson distribution with parameter $\mu(A)$:

$$\mathbb{P}(N(A) = k) = e^{-\mu(A)} \frac{\mu(A)^k}{k!}, \quad k = 0, 1, \ldots.$$

2. For every countable family of Borel sets $(A_i)_{i \in \mathcal{I}}$ with $A_i \cap A_j = \emptyset$, if $i \neq j$, the random variables $\big(N(A_i)\big)_{i \in \mathcal{I}}$ are independent.

Consider now a random measure N on $\mathcal{X} = [0, \infty) \times [0, \infty)$ with intensity $dt \otimes dx$. A first natural process that can be easily constructed from N is a so-called inhomogeneous Poisson process with intensity function $\lambda : [0, \infty) \to [0, \infty)$. It models an event based process $(X_t)_{t \geq 0}$ that counts the events that occur between 0 and t according to the following two rules:

$\mathbb{P}(\text{an event occurs in } (t, t + dt) \mid \text{given the history up to time } t) = \lambda(t)dt,$

and $(X_t)_{t\geq 0}$ has independent increments: the random variables: $(X_{t_i+h_i} - X_{t_i})_{1\leq i\leq n}$ are independent, for any family of disjoints sets $([t_i, t_i+h_i])_{1\leq i\leq n}$. A possible construction is given by

$$X_t = \int_0^t \int_{[0,\infty)} \mathbb{1}_{\{u\leq\lambda(s)\}} N(ds, du).$$

To better understand the construction from a heuristic point of view, if $\lambda(s)$ is sufficiently smooth, for $s \in [t, t+dt]$, we can approximate $s \mapsto \lambda(s)$ by the constant function $s \mapsto \lambda(t)$ and thus obtain the chain of approximations

$$X_{t+dt} - X_t = \int_t^{t+dt} \int_0^{\lambda(s)} N(ds, du)$$

$$\approx \int_t^{t+dt} \int_0^{\lambda(t)} N(ds, du) = N\big([t, t+dt] \times [0, \lambda(t)]\big)$$

which has a Poisson distribution of parameter $\lambda(t)dt$ since the intensity of N is $dt \otimes ds$. In particular, a standard Poisson process with parameter $\lambda > 0$ can be represented as $X_t = \int_0^t \int_{[0,\infty)} \mathbb{1}_{\{u\leq\lambda\}} N(ds, du)$. This is the basic ingredient to construct more elaborate jump models.

The simplest age model

We first prove the link between the deterministic and the stochastic model in the simplest age-dependent model: at time t, the age-structured cell population is described by the states (here the ages) of the living cells at time t, that we denote $(A_1(t), A_2(t), \ldots, A_n(t), \ldots)$ (given in the increasing order for instance), that we encode into the point measure

$$Z_t^{(a)} = Z_t(da) = \sum_{i=1}^{\infty} \delta_{A_i(t)}(da),$$

where the sum ranges from 1 to $\langle Z_t, \mathbb{1}\rangle$ and which is finite, as in (2.2). Assuming that the ages are ordered, we define the evaluation maps

$$a_i : Z_t \mapsto a_i(Z_t) = a_i\left(\sum_{j=1}^{\infty} \delta_{A_j(t)}(da)\right) = A_i(t).$$

Abusing notation slightly, we may identify $A_i(t)$ and $a_i(Z_t)$. We have a complete description of the stochastic dynamics of $Z_t = Z_t^{(k)}$ by means of a family of independent Poisson random measures $N_i(ds, d\vartheta)$, $i = 1, 2 \ldots$ with intensity $ds \otimes d\vartheta$ on $[0, \infty) \times [0, \infty)$. It is given by the stochastic evolution equations, for $k = 1, 2$:

$$Z_t^{(k)} = \tau_t Z_0$$

$$+ \int_0^t \sum_{i \leq \langle Z_{s-}^{(k)}, \mathbb{1} \rangle} \int_0^\infty (k\delta_{t-s} - \delta_{a_i(Z_{s-}^{(k)})+t-s}) \mathbb{1}_{\{\vartheta \leq B(a_i(Z_{s-}^{(k)}))\}} N_i(ds, d\vartheta), \tag{2.3}$$

where

$$Z_0^{(k)} = \sum_{i=1}^{\langle Z_0^{(k)}, \mathbb{1} \rangle} \delta_{A_i(0)}$$

is an initial age distribution and

$$\tau_t Z_0^{(k)} = \sum_{i=1}^{\langle Z_0^{(k)}, \mathbb{1} \rangle} \delta_{A_i(0)+t}.$$

We have $k = 1$ for the genealogical observation case, where only one daughter cell is kept at each division, and $k = 2$ for the population observation case.

We may interpret (2.3) as follows: at time t the term $\tau_t Z_0^{(k)}$ accounts for the initial population that has aged by the amount t. This term must be corrected by:

- adding the ages of newborn cells. This is done as follows: a division event is run according to the rate $B(a_i(Z_{s-}^{(k)}))$, where i ranges over the whole population at time s, (and then s ranges from 0 to t). When a division occurs, produced by the cell i with age $a_i(Z_{s-}^{(k)})$, k newborn cells are created with age 0 at time s, that will have the same age $t - s$ at time t. We thus add to the system the term $k\delta_{t-s}$;
- removing the ages of the mother cells. This comes from the dying mother cell at time s and age $a_i(Z_{s-}^{(k)})$, according to the same division event as for the addition of newborn cells. This cell would have age $a_i(Z_{s-}^{(k)})+t-s$ at time t, and therefore, we remove to the system the term $\delta_{a_i(Z_{s-}^{(k)})+t-s}$.

Whenever the integrator is a jump measure, we must be careful to consider integrands that are predictable in the sense of stochastic calculus, hence the left limits in terms involving $a_i(Z_{s-})$. This is innocuous at the informal level we keep here, but becomes crucial whenever martingales properties are involved, that are essential to properly define existence and uniqueness of a measure solution $(Z_t^{(k)})_{t\geq 0}$ to (2.3).

Note that it is not completely obvious that a solution $(Z_t^{(k)})_{t\geq 0}$ to (2.3) exists, as well as its uniqueness. This is obtained by classical arguments and we refer to Refs. [41, 42] for a comprehensive treatment of the subject.

From the stochastic evolution (2.3) to a deterministic PDE

Consider the following renewal equation:

$$\begin{cases} \frac{\partial}{\partial t} n_k(t,a) + \frac{\partial}{\partial a} n_k(t,a) + B(a)n_k(t,a) = 0, \\ \\ n_k(t,0) = k \int_0^\infty B(a)n_k(t,a)da, \\ \\ n_k(0,a) = n^{in}(a), \qquad \int_0^\infty n^{in}(a)da = 1. \end{cases} \qquad (2.4)$$

This is a classical model that goes back to McKendricks and von Foerster, see in particular the textbooks [34, 46]. If n^{in} and B are well-behaved there is existence and weak uniqueness of (2.4), see Proposition 3.3. Moreover, the solution $n_k(t,da) = n_k(t,a)da$ is absolutely continuous. Define for $t \geq 0$ the deterministic family of positive measures $(m_k(t))_{t\geq 0}$ via

$$\langle m_k(t,\cdot), \varphi \rangle = \int_0^\infty \varphi(a)m_k(t,da) = \mathbb{E}\big[\langle Z_t^{(k)}, \varphi \rangle\big],$$

with $m_k(0,da) = n^{in}(a)da$. The interpretation of m is the mean, or macroscopic state of the system. The link between the stochastic evolution (2.3) and the deterministic Fokker-Planck equation (2.4) is given by the following simple equivalence:

Proposition 2.2: *Assume that B is continuous and bounded and that the initial condition n^{in} is absolutely continuous, with a bounded continuous density. We then have $m_k = n_k$.*

These conditions are not minimal. We detail the essential steps of the proof of Proposition 2.2, since it illustrates in a relatively simple framework the interplay between deterministic and stochastic methods. The other equivalence results between random measure evolution equations and structured population equations stated in Sec. 2.3 are obtained in the same way.

Proof: *Step 1) The action of time and age test functions (2.3).* Let $\varphi_t(a) = \varphi(t,a)$ denote a test function (assumed to be smooth and compactly supported for safety). First, by definition of (2.3),

$$\langle Z_t^{(k)}, \varphi_t \rangle = \langle \tau_t Z_0^{(k)}, \varphi_t \rangle + \int_0^t \sum_{i \le \langle Z_{s-}^{(k)}, \mathbb{1} \rangle} \int_0^\infty$$

$$\left(k\varphi_t(t-s) - \varphi_t(a_i(Z_{s-}^{(k)}) + t - s) \right) \mathbb{1}_{\{\vartheta \le B(a_i(Z_{s-}^{(k)}))\}} N_i(ds, d\vartheta). \tag{2.5}$$

Next, for every $0 \le s \le t$, we have

$$\varphi_t(a + t - s) = \varphi_s(a) + \int_s^t \left(\tfrac{\partial}{\partial u}\varphi_u(a + s - u) + \tfrac{\partial}{\partial a}\varphi_u(a + s - u) \right) du.$$

Therefore, successively

$$\langle \tau_t Z_0^{(k)}, \varphi_t \rangle = \sum_{i=1}^{\langle Z_0^{(k)}, \mathbb{1} \rangle} \left(\varphi_0(a_i(Z_0^{(k)})) \right.$$

$$\left. + \int_0^t \left(\tfrac{\partial}{\partial s}\varphi_s(a_i(Z_0^{()k}) + s) + \tfrac{\partial}{\partial a}\varphi_s(a_i(Z_0^{(k)}) + s) \right) ds \right),$$

$$\varphi_t(t-s) = \varphi_s(0) + \int_s^t \left(\tfrac{\partial}{\partial u}\varphi_u(u-s) + \tfrac{\partial}{\partial a}\varphi_u(u-s) \right) du,$$

$$\varphi_t(a_i(Z_{s-}^{(k)}) + t - s) = \varphi_s(a_i(Z_{s-}^{(k)})) +$$

$$\int_s^t \left(\tfrac{\partial}{\partial u}\varphi_u(a_i(Z_{s-}^{(k)}) + u - s) \tfrac{\partial}{\partial a}\varphi_u(a_i(Z_{s-}^{(k)}) + u - s) \right) du.$$

Plugging-in these three expansions in (2.5), and using the definition of $Z_s^{(k)}$ again, we obtain

$$\langle Z_t^{(k)}, \varphi_t \rangle = \langle Z_0^{(k)}, \varphi_0 \rangle + \int_0^t \langle \tfrac{\partial}{\partial s}\varphi_s + \tfrac{\partial}{\partial a}\varphi_s, Z_s^{(k)} \rangle ds$$

$$+ \int_0^t \sum_{i \le \langle Z_{s-}^{(k)}, \mathbb{1} \rangle} \int_0^\infty \left(k\varphi_s(0) + \varphi_s(a_i(Z_{s-}^{(k)})) \right) \mathbb{1}_{\{\vartheta \le B(a_i(Z_{s-}^{(k)}))\}} N_i(ds, d\vartheta).$$

Step 2) A martingale-oriented representation of (2.3). We compensate the Poisson random measures $N_i()ds, d\vartheta$ and define

$$\widetilde{N}_i(ds, d\vartheta) = N(ds, d\vartheta) - ds \otimes d\vartheta.$$

For fixed φ, by construction, the random process

$$M_t(\varphi) = \int_0^t \sum_{i \le \langle Z_{s-}^{(k)}, \mathbb{1} \rangle} \int_0^\infty \left(k\varphi_s(0) + \varphi_s(a_i(Z_{s-}^{(k)})) \right) \mathbb{1}_{\{\vartheta \le B(a_i(Z_{s-}^{(k)}))\}} \widetilde{N}_i(ds, d\vartheta)$$

is a centred martingale, and in particular $\mathbb{E}[M_t(\varphi)] = 0$. Noticing that

$$\sum_{i \leq \langle Z_{s-}^{(k)}, \mathbb{1} \rangle} \int_0^\infty (k\varphi_s(0) + \varphi_s(a_i(Z_{s-}^{(k)}))) \mathbb{1}_{\{\vartheta \leq B(a_i(Z_{s-}^{(k)}))\}} ds d\vartheta$$

$$= \langle Z_s^{(k)}, (k\varphi_s(0) - \varphi_s)B \rangle,$$

we obtain from Step 1 the representation

$$\langle Z_t^{(k)}, \varphi_t \rangle = \langle Z_0^{(k)}, \varphi_0 \rangle + \int_0^t \langle \tfrac{\partial}{\partial s}\varphi_s + \tfrac{\partial}{\partial a}\varphi_s + (k\varphi_s(0) - \varphi_s)B, Z_s^{(k)} \rangle ds + M_t(\varphi).$$

Taking expectation and applying Fubini's theorem, we derive

$$\langle m_k(t, \cdot), \varphi_t \rangle = \langle m_k(0, \cdot), \varphi_0 \rangle + \int_0^t \langle \tfrac{\partial}{\partial s}\varphi_s + \tfrac{\partial}{\partial a}\varphi_s + (k\varphi_s(0) - \varphi_s)B, m_k(s, \cdot) \rangle ds.$$

$$(2.6)$$

Step 3) From (2.6) to (2.4). We finally prove that (2.6) and (2.4) are equivalent, along a classical line of arguments: we take for granted that $m_k(t, da) = m_k(t, a)da$ is absolutely continuous. This is a consequence of the smoothness of the initial condition and of B, see Proposition 3.3. We refer to Ref. [34] for transport equations or to Ref. [41] for a probabilist point of view. We evaluate its time derivative against a test function ϕ that depends on age only using (2.6). We obtain, for a nice enough ϕ so that we can interchange integral and derivative:

$$\tfrac{d}{dt} \int_0^\infty \phi(a)m_k(t, a)da = \int_0^\infty (\tfrac{\partial}{\partial a}\phi(a) + (k\phi(0) - \phi(a))B(a))m_k(t, a)da.$$

Assume now that ϕ is compactly supported and smooth, and that moreover $\phi(0) = 0$. Integrating by part

$$\int_0^\infty \tfrac{\partial}{\partial a}\phi(a)m_k(t, a)da = -\int_0^\infty \phi(a)\tfrac{\partial}{\partial a}m_k(t, a)da.$$

We infer that for every such ϕ: $\tfrac{d}{dt}\langle m_k(t, \cdot), \phi \rangle = -\int_0^\infty (\tfrac{\partial}{\partial a}m_k(t, a) + B(a))m_k(t, a)\phi(a)da$ and therefore

$$\tfrac{\partial}{\partial t}m_k(t, a) + \tfrac{\partial}{\partial a}m_k(t, a) + B(a)m(t, a) = 0 \quad da - \text{almost everywhere}$$

using one more time that $\phi(0) = 0$ to eliminate the term involving $k\phi(0)$. Now, consider an arbitrary test function ϕ vanishing at infinity: from the previous computation, we have

$$0 = \int_0^\infty \left(\tfrac{\partial}{\partial t}m_k(t, a) + \tfrac{\partial}{\partial a}m_k(t, a) \right) \phi(a)da$$

$$= -\phi(0)m_k(t, 0) + k\phi(0) \int_0^\infty m_k(t, a)B(a)da$$

and simplifying, we obtain the boundary condition $m_k(t, 0) = k \int_0^\infty B(a) m_k(t, a)$, as expected. This completes the proof. □

2.3. *Structured population equations*

In the previous subsection, we have seen a rigorous derivation of the renewal equation from the stochastic differential equation. The same proof can be done in more intricate situations. For specific examples, we now give below the stochastic equation followed by the corresponding PDE, and the references for their rigorous derivation as well as some examples of possible generalisations.

Size-structured model: the growth-fragmentation process & equation

We structure the cell population according to the individual sizes $X_i(t)$ of each individual present at time t, encoded into the random point measure

$$Z_t^{(k)}(dx) = \sum_{i=1}^\infty \delta_{X_i(t)}(dx),$$

with values in $(0, \infty)$, the state space of sizes, and $k = 1$ or 2 refers to the number of cells that are kept into the system at division. The sum ranges from 1 to $\langle Z_t^{(k)}, \mathbb{1} \rangle$. Assuming that the sizes are ordered, the evaluation maps are defined as

$$x_i(Z_t^{(k)}) = x_i \left(\sum_{j=1}^\infty \delta_{X_j(t)}(dx) \right) = X_i(t).$$

In analogy to the age model described in the previous section, we have a stochastic evolution equation for the process $(Z_t^{(k)})_{t \geq 0}$:

$$Z_t^{(k)} = \exp(\kappa t) Z_0 + \int_0^t \sum_{i \leq \langle Z_{s-}^{(k)}, \mathbb{1} \rangle} \int_0^\infty$$

$$(k \delta_{\frac{1}{2} x_i(Z_{s-}^{(k)}) \exp(\kappa(t-s))} - \delta_{x_i(Z_{s-}^{(k)}) \exp(\kappa(t-s))}) \mathbb{1}_{\{\vartheta \leq B(x_i(Z_{s-}^{(k)}))\}} N_i(ds, d\vartheta),$$

(2.7)

where, abusing notation slightly, for a random point measure $Z(dx) = \sum_{i=1}^\infty \delta_{z_i}(dx)$, and a real valued function ϕ, we set $\phi(Z)(dx) = \sum_{i=1}^\infty \delta_{\phi(z_i)}(dx)$, so that $\exp(\kappa t) Z_0 = \sum \delta_{X_i(0) e^{\kappa t}}$. Set, for (regular compactly supported) φ

$$\langle n_k(t, \cdot), \varphi \rangle = \mathbb{E} \left[\langle Z_t, \varphi \rangle \right].$$

We have (in a weak sense)

$$\frac{\partial}{\partial t} n_k(t, x) + \frac{\partial}{\partial x}\big(\kappa x\, n_k(t, x)\big) + B(x) n_k(t, x) = 2kB(2x) n_k(t, 2x). \quad (2.8)$$

The fact that the measure $\mathbb{E}\big[\langle Z_t, \cdot \rangle\big]$ solves (2.8) can be obtained along the same line or arguments as in Proposition 2.2. An alternative proof related on fragmentation processes, following the tagged fragment technique developed for instance in Refs. [43, 44], can be found in Ref. [37], for a more general model allowing variability in the growth rate κ.

Adder model: the incremental process & structured equation

We structure the model in the pair of traits (x, z) where x denotes the size of a cell and z its size increment since its birth. We obtain a random measure

$$Z_t^{\iota}(dx, dz) = \sum_{i=1}^{\infty} \delta_{(X_i(t), Z_i(t))}(dx, dz),$$

where $(X_1(t), X_2(t) \ldots)$ denotes the (ordered) size of each cell present in the system at time t (being born before t of course) and $Z_i(t)$ denotes the size increment of the cell with size X_i^b at birth. With these notation, the size $X_i(t)$ of a cell present in the system at time t is simply

$$X_i(t) = X_i^b + Z_i(t).$$

With the evaluation mappings

$$(x_i, z_i)(Z_t^{\iota}) = (x_i, z_i)\left(\sum_{j=1}^{\infty} \delta_{(X_j(t), Z_j(t))}(dx, dx)\right) = (X_i(t), Z_i(t)),$$

the stochastic evolution for the measure-valued process $(Z_t^{\iota})_{t \geq 0}$ is given by

$$Z_t^{\iota} = Z_0^{\iota} \exp(\kappa t)$$
$$+ \int_0^t \sum_{i \leq \langle Z_{s-}^{\iota}, \mathbb{1} \rangle} \int_0^{\infty} \Big(k \delta_{\big(\frac{1}{2} x_i(Z_{s-}^{\iota}) \exp(\kappa(t-s)), \frac{1}{2} x_i(Z_{s-}^{\iota})(e^{\kappa(t-s)} - 1)\big)}$$
$$- \delta_{\big(x_i(Z_{s-}^{\iota}) e^{\kappa(t-s)}, x_i(Z_{s-}^{\iota}) e^{\kappa(t-s)} - (x_i(Z_{s-}^{\iota}) - z_i(Z_{s-}^{\iota}))\big)} \Big)$$
$$\times \mathbb{1}_{\{\vartheta \leq \kappa x_i(Z_{s-}^{\iota}) B(z_i(Z_{s-}^{\iota}))\}} N_i(ds, d\vartheta), \quad (2.9)$$

where

$$Z_0^{\iota} \exp(\kappa t) = \sum_{i=1}^{\infty} \delta_{\big(x_i(Z_0^{\iota}) \exp(\kappa t), x_i(Z_0^{\iota}) \exp(\kappa t) - (x_i(Z_0^{\iota}) - z_i(Z_0^{\iota}))\big)}(dx, dz),$$

and similarly, set for (regular compactly supported) φ valued in $(0, \infty) \times (0, \infty)$

$$\langle n_k(t, \cdot, \cdot), \varphi \rangle := \mathbb{E}\left[\langle Z_t, \varphi \rangle\right].$$

We have (in a weak sense)

$$\frac{\partial}{\partial t} n_k + \frac{\partial}{\partial z}(\kappa x n_k) + \frac{\partial}{\partial x}(\kappa x n_k) = -\kappa x B(z) n_k(t, z, x), \qquad (2.10)$$

$$\kappa x n_k(t, 0, x) = 4k\kappa x \int_0^\infty B(z) n_k(t, z, 2x) dz, \qquad (2.11)$$

with $n_k(0, z, x) = n^{(0)}(z, x)$, $0 \le z \le x$.

Discussion on some generalisations

We can follow two variables, one behaving like a physiological age, *i.e.* reset at zero at each division, but which evolves with a non-constant rate $\tau_z(z, x)$; the other behaving like a size, *i.e.* which is conserved by division through a fragmentation kernel $b(y, dx)$ such that $\int_0^y b(y, dx) = 1$, and growing at a rate $\tau_x(x, \kappa)$, with a parameter κ chosen at birth according to the rate κ' of the mother along a probability kernel $\theta(\kappa', d\kappa)$. We may assume that the division rate depends on size, growth rate and physiological age, and denote it $\beta(z, x, \kappa)$; we write it as a time instantaneous rate, contrarily to the previous notations where B is an age or a size instantaneous rate, thus multiplied by the corresponding age or size growth rate. We expect the mean measure $n_k = \mathbb{E}[\langle Z_t, \cdot \rangle]$ of the corresponding population process to solve a PDE of the form

$$\frac{\partial}{\partial t} n_k + \frac{\partial}{\partial z}(\tau_z(z, x) n_k) + \frac{\partial}{\partial x}(\tau_x(x, \kappa) n_k) = -\beta(z, x, \kappa) n_k(t, z, x, \kappa),$$
$$(2.12)$$

$$\tau_x(0, \kappa) n_k(z, 0, \kappa) = 0, \qquad \tau_z(0, \kappa) n_k(t, 0, x, \kappa) = \qquad (2.13)$$

$$2k \int_0^\infty \int_0^\infty \int_0^\infty \theta(\kappa', \kappa) b(y, x) \beta(z, y, \kappa') n_k(t, z, y, \kappa') dz dy d\kappa'.$$

We recover the age model by taking $\beta(z, x, \kappa) = B(z)$, $\tau_z \equiv 1$, and integrating in size and growth rate; the size model with variable growth rate and unequal fragmentation by taking $\beta(z, x, \kappa) = \tau_x(x, \kappa) B(x)$ and integrating in z; the increment model with variable growth rate by taking $\tau_z = \tau_x$,

$\beta(z, x, \kappa) = \tau_x(x, \kappa)B(z)$. A detailed study lies beyond the level intended in these notes but the interest is to embed all the models considered here in a common framework.

3. Model analysis: long-time behaviour

Having built kinetic models leads naturally mathematicians towards the question of their long-time behaviour. In this matter, this is far from being a pure mathematical question: it reveals the cornerstone of our calibration strategy, developed in Sec. 4. It is however a whole field in itself: as in Sec. 2, we provide the main ingredients in the simplest case of the renewal equation and process — which already reveals not so simple when we study the stochastic population model — and review — not exhaustively — the extremely rich literature for more involved cases.

Importantly, we only consider here linear models, *i.e.* we neglect feedbacks or exchanges with the environment or between the cells; we only mention a few nonlinear results. When linear, the study of the asymptotic behaviour of the equations are closely related to the spectral analysis of the related semigroup operator, and leads to three main types of behaviours:

- convergence to a steady state (exponentially fast in case of a spectral gap), which happens for the conservative equations (case $k = 1$: genealogical observation);
- steady exponential growth, *i.e.* there is a decoupling between an exponential growth at the rate of the dominant eigenvalue, called the *Malthusian parameter* or *fitness* of the population, and a steady distribution in the structuring variables. This happens in the case $k = 2$, where all the population is followed. As for the conservative case, an exponentially-fast trend to this steady behaviour is linked to a spectral gap.
- Other non-physical behaviours, for instance spreading in the distributions, or yet trend to a permanently oscillating system, together with exponential growth in the case $k = 2$. This last case is a non-robust behaviour, linked to some degeneracy in the coefficients, so that the dissipation of entropy is not sufficient for a trend to a steady behaviour to emerge; it can also be seen as another type of convergence towards the dominant eigenvector, when this one becomes non-unique. Whatsoever, this last behaviour, in the case of linear equations, is a mathematical curiosity rather than biologically informative. The interest of proving such non-physical results consists in excluding some models or assumptions as non-realistic if they lead to such non-physical results.

These considerations drive our assumptions: they need to ensure the convergence of both the conservative ($k = 1$) and the non-conservative ($k = 2$) equations towards steady behaviours.

3.1. *The renewal equation: a pedagogical example*

This is historically the first structured-population model to be studied [45,46]; another key reference is [26] around 1950 and later the celebrated textbook book by Harris [47]. There are strong links with the classical renewal theory of random walks in probability, that dates back to classics: W. Feller, J. Doob, A. Lotka [48–50]. A recent account of the link between fragmentation processes and renewal theory can be found in the textbook by J. Bertoin [43].

Let us denote the renewal equation in one of its simplest form:

$$\begin{cases} \frac{\partial}{\partial t} n(t,a) + \frac{\partial}{\partial a} n(t,a) = -B(a)n(t,a), \\ \\ n(t,0) = k \int_0^\infty B(a)n(t,a)da, \qquad n(0,a) = n^0(a). \end{cases} \tag{3.1}$$

This equation has to be understood first in a weak sense, *i.e.* as an equivalent way to write (2.4).

Functional spaces through the lense of modelling

To give a rigorous meaning to this equation, either in a weak or strong formulation, one first needs to decide in which functional space — in the structuring variable (age here), then time — the solution n should be considered.

Closer to the stochastic process and to non-asymptotic or not-averaged populations are measure-valued solutions, see the recent and very pedagogical approach of P. Gabriel [51] for the case $k = 1$, or [52]: this allows one to consider measure-valued initial conditions for n^0.

Convenient to handle the inverse problem or to use Fourier or Mellin transforms are L^2- types spaces, see e.g. Ref. [12], but they lack physical interpretation. Assuming however that biological populations are expected to be in both L^1 and L^∞, this is acceptable, for mathematical reasons, to choose such spaces.

Finally, weighted L^1 spaces have been widely used, for semi-group approaches [46] as well as for general relative entropy [34] or still others [53]. They have the advantage of immediate physical interpretation, since $\int n(t,a)da$ and $\int B(a)n(t,a)da$ represent respectively (the expectation of) the total number of individuals and the total number of dividing individuals at time t, $\int an(t,a)da$ the average age of individuals at time t, etc. Moreover, assuming $n(t,\cdot) \in L^1((0,\infty))$ means that the expectation of the stochastic measure has a density and, up to a renormalization if $k = 2$, is a probability density. At least for large times, we consider that this is relevant in a modelling perspective, and thus privilege this last family of functional spaces.

Let us briefly recall how the main results may be found using the entropy approach developed in Ref. [54], and discuss the links between the cases $k = 0$ and $k = 1$ — we refer to Ref. [34], Chapter 2 for a complete presentation. We assume

$$B \in L_{loc}^\infty \left((0,\infty), [0,\infty)\right),$$

$$\int^\infty B(x)dx = \infty, \qquad \int_0^\infty xB(x)e^{-\int_0^x B(a)da}dx < \infty.$$

(3.2)

These assumptions have a stochastic interpretation: if A is a random variable representing the age at division of a cell, the fact that $\int_0^\infty B(x)dx = \infty$ ensures that all cells divide. The probability density of A being the function $x \mapsto B(x)e^{-\int_0^x B(a)da}$, the last assumption means that $\mathbb{E}[A] < \infty$.

Eigenelements

As said in the introduction, the asymptotic behaviour is closely linked to the study of the spectrum of the linear operator under consideration: we may formally write

$$\frac{d}{dt}n(t,\cdot) = \mathcal{A}_k n(t,\cdot)$$

where $\mathcal{A}_k$ is defined by (3.1). Eigenelements $(\lambda_k, N_k, \varphi_k)$ are solutions to the equations

$$\mathcal{A}_k N_k = \lambda_k N_k, \qquad \mathcal{A}_k^* \varphi_k = \lambda_k \varphi_k,$$

where $\mathcal{A}_k^*$ is the adjoint operator to $\mathcal{A}_k$, defined by the following adjoint equation

$$\mathcal{A}_k^*\varphi(a) := \frac{\partial}{\partial a}\varphi(a) - B(a)\varphi(a) + kB(a)\varphi(0).$$

Proposition 3.1: *Under Assumption* (3.2), *for $k = 1$ and $k = 2$, there* *exists a unique solution* $(\lambda_k, N_k, \varphi_k)$ *to the system*

$$
\begin{cases}
\lambda_k N_k(a) + \frac{\partial}{\partial a} N_k(a) + B(a) N_k(a) = 0, \qquad \int_0^\infty N_k(a) da = 1, \\[2mm]
N_k(0) = k \int_0^\infty B(a) N_k(a) da, \\[2mm]
\lambda_k \varphi_k(a) - \frac{\partial}{\partial a} \varphi_k(a) + B(a) \varphi_k(a) = k B(a) \varphi_k(0), \int_0^\infty N_k(a) \varphi_k(a) da = 1.
\end{cases}
$$
$$\tag{3.3}$$

Moreover, we have $\lambda_1 = 0$, $\varphi_1 \equiv 1$, $\lambda_2 > 0$, N_k, $\varphi_2 > 0$, λ_2 *is uniquely defined by the relation*

$$
1 = 2 \int_0^\infty B(a) e^{-\lambda_2 a - \int_0^a B(s) ds} da,
\tag{3.4}
$$

and $\varphi_2 \in L^\infty(0, \infty)$, $\varphi_2(0) > 0$ *with the uniform bound*

$$
\|\varphi_2\|_{L^\infty} \leq 2\varphi_2(0) = 2 \frac{\int_0^\infty e^{-\lambda_2 a - \int_0^a B(s) ds} da}{\int_0^\infty s B(s) e^{-\lambda_2 s - \int_0^s B(\sigma) d\sigma} ds}.
$$

Proof: The fact that $\varphi_1 \equiv 1$ can be seen directly from the fact that for $k = 1$ the equation is conservative. In the case of the renewal equation, contrarily to more involved cases where the study of existence and uniqueness of eigenelements is a field in itself (see Sec. 3.3), we can immediately compute that solutions must satisfy

$$
\begin{cases}
N_k(a) = N_k(0) e^{-\lambda_k a - \int_0^a B(s) ds}, N_k(0) = N_k(0) k \int_0^\infty B(a) e^{-\lambda_k a - \int_0^a B(s) ds} da, \\[2mm]
\varphi_k(a) = \varphi_k(0) \left(1 - k \int_0^a B(s) e^{-\lambda_k - \int_0^s B(\sigma) d\sigma} ds\right) e^{\lambda_k a + \int_0^a B(s) ds} \\[2mm]
\qquad = k \varphi_k(0) \int_a^\infty B(s) e^{-\lambda_k(s-a) - \int_a^s B(\sigma) d\sigma} ds.
\end{cases}
$$
$$\tag{3.5}$$

From this relation and from Assumption (3.2), we deduce that $\lambda_1 = 0$ and obtain that λ_2 satisfies (3.4) from the boundary condition at $a = 0$. Since the right-hand side of (3.4) is a continuously decreasing function of λ_2 that equals 2 for $\lambda_2 = 0$ (thanks to the fact that $\int_0^\infty B dx = \infty$) and vanishes for $\lambda \to \infty$, this defines a unique $\lambda_2 > 0$. The normalisation condition $\int_0^\infty N_k(a) da = 1$ ensures uniqueness of N_k (and the convergence of the integral $\int_0^\infty N_1(a) da = 1$ is guaranteed by the assumption $\int x B(x) e^{\int_0^x B da} < \infty$); the normalisation condition $\int_0^\infty N_k(a) \varphi_k(a) da = 1$ ensures uniqueness of φ_k. The uniform bound for φ_2 is obtained by integrating by parts (3.5):

$$\varphi_2(a) = 2\varphi_2(0)\left(1 - \int_a^\infty \lambda_2 e^{-\lambda_2(s-a)-\int_a^s B(\sigma)d\sigma} ds\right) \le 2\varphi_2(0),$$

and we compute

$$1 = \int_0^\infty N_2(a)da \implies N_2(0) = \left(\int_0^\infty e^{-\lambda_2 a - \int_0^a B(s)ds} da\right)^{-1},$$

$$1 = \int_0^\infty N_2(a)\varphi_2(a)da = N_2(0)\varphi_2(0)\int_0^\infty \int_a^\infty B(s)e^{-\lambda s - \int_0^s B(\sigma)d\sigma} ds\,da$$

$$= N_2(a)\varphi_2(a)\int_0^\infty sB(s)e^{-\lambda s - \int_0^s B(\sigma)d\sigma} ds$$

$$\implies \varphi_2(0) = \frac{\int_0^\infty e^{-\lambda_2 a - \int_0^a B(s)ds} da}{\int_0^\infty sB(s)e^{-\lambda_2 s - \int_0^s B(\sigma)d\sigma} ds}. \qquad \square$$

Remark 3.2: We may find a solution for $k = 2$ under the relaxed assumption

$$B \in L_{loc}^\infty\left((0,\infty),[0,\infty)\right), \quad \int_0^\infty B(x)dx > \ln(2),$$

which is weaker than (3.2). This would however lead to $\lambda_1 < 0$ in the case where $\int_0^\infty Bdx < \infty$, so that the conservative equation would have no conservative eigenvector, *i.e.* no possible trend to a steady behaviour, what should be excluded for modelling purpose. Similarly, if the assumption $\int xB(x)e^{-\int_0^x Bda}dx < \infty$ is not fulfilled, we cannot normalise the eigenpair by $\int N_1\varphi_1 dx = 1$ because $\int N_1\varphi_1 dx = \infty$ in this case.

Existence of solutions

Many methods are available to prove existence and uniqueness of solutions in various spaces; we specially refer the interested reader to Refs. [34, 46] and for measure solutions to Ref. [51]. We only mention here an easy proof obtained by the Banach-Picard fixed point theorem and Duhamel's formula, inspired by Ref. [34], Chapter 3, [55] Appendix B and [51] (solution for the dual equation). We take an L^1 space which is natural for solutions, namely $L^1\left((1+B(a))da\right)$, so that the assumptions on B are minimal.

Proposition 3.3: *For $B \in L_{loc}^\infty\left((0,\infty);(0,\infty)\right)$ such that $\int_0^\infty B(a)da = \infty$, for $n^0 \in L^1\left((1+B(a))da\right)$, there exists a unique solution $n \in L_{loc}^\infty\left(0,\infty;L^1\left((1+B(a))da\right)\right)$ to (3.1) and we have the comparison principle:*

$$\forall x, \, n_1^0(a) \le n_2^0(a) \implies n_1(t,a) \le n_2(t,a) \qquad \forall \, t, \, a \ge 0.$$

Proof: Let $\lambda > 0$ a given constant. We look for solutions $\tilde{n}(t,a) = e^{-\lambda t}n(t,a)$ to the equation

$$\begin{cases} \frac{\partial}{\partial t}\tilde{n}(t,a) + \frac{\partial}{\partial a}\tilde{n}(t,a) + \lambda\tilde{n} = -B(a)\tilde{n}(t,a), \\ \tilde{n}(t,0) = k\int_0^\infty B(a)\tilde{n}(t,a)da, \qquad \tilde{n}(0,a) = n^0(a). \end{cases}$$

We define $\psi(t) = \tilde{n}(t,t+a)$ which is solution to

$$\psi'(t) + (B(t+a)+\lambda)\,\psi = 0, \qquad \psi(0) = n^0(a),$$

and similarly $\tilde{\psi}(t) = \tilde{n}(t+a,t)$ is solution to

$$\tilde{\psi}' + (B(t)+\lambda)\,\tilde{\psi} = 0, \qquad \tilde{\psi}(0) = \tilde{n}(a,0) = k\int_0^\infty B(s)\tilde{n}(a,s)ds$$

so that we find the Duhamel's formula

$$\tilde{n}(t,a) = n^0(a-t)e^{-\int_0^t B(s+a)ds-\lambda t}\mathbb{1}_{\{a\geq t\}}$$

$$+ke^{-\int_0^a B(\sigma)d\sigma-\lambda a}\int_0^\infty B(s)\tilde{n}(t-a,s)ds\mathbb{1}_{\{t\geq a\}} \qquad (3.6)$$

$$:= \Gamma_{n^0}[\tilde{n}](t,a).$$

We consider the mapping Γ_{n^0} taken on the weighted space $X = L^\infty\left(0,\infty; L^1\left((\frac{\lambda}{2k}+B(a))\,da\right)\right)$: we find, for any $t \geq 0$,

$$\int_0^\infty \left|\left(\tfrac{\lambda}{2k}+B(a)\right)\Gamma[n_1-n_2](t,a)\right|da$$

$$\leq \int_0^\infty k\left(\tfrac{\lambda}{2k}+B(a)\right)e^{-\int_0^a B(\sigma)d\sigma-\lambda a}da\,\|n_1-n_2\|_X$$

$$\leq k\left(1+\int_0^\infty(\tfrac{\lambda}{2k}-\lambda)e^{-\lambda a}da\right)\|n_1-n_2\|_X$$

$$\leq \tfrac{1}{2}\|n_1-n_2\|_X.$$

This proves that Γ is a strict contraction on X, hence applying the Banach-Picard fixed point theorem we find a unique fixed point, which is solution to (3.1). The comparison principle comes from the fact that if $n_0^1 \leq n_0^2$ then $\Gamma_{n_1^0}[n] \leq \Gamma_{n_2^0}[n]$; we can iterate each of the operators $\Gamma_{n_i^0}$ and at the limit we find $\tilde{n}_1 \leq \tilde{n}_2$, hence $n_1 \leq n_2$. $\qquad\square$

General relative entropy

A general fact shared by many linear population dynamics equations is a wide class of time-decreasing functionals, which may be used as a key ingredient to study the equation, in particular to prove long-time asymptotics [35] or uniqueness of eigensolutions [36]. We refer to Ref. [34] for a thorough presentation in many situations.

Lemma 3.4: *Let $H : [0, \infty) \to [0, \infty)$ a convex differentiable function, and n_1, n_2 two solutions of (3.1) in $L^{\infty}_{loc}([0, \infty); L^1((1+B(a))da)$ such that $n_2(t, x) > 0$ for all $t, x \geq 0$, and such that $\int \varphi_k(a)n_2(0, a)H\left(\frac{n_1(0,a)}{n_2(0,a)}\right) da < \infty$. We have the following inequality:*

$$\mathcal{H}(t) := \int \varphi_k(a)e^{-\lambda_k t}n_2(t, a)H\left(\frac{n_1(t, a)}{n_2(t, a)}\right) da \implies \frac{d}{dt}\mathcal{H}(t) \leq 0.$$
(3.7)

Proof: We compute, using the equations and integrating by parts:

$$\frac{d}{dt}\mathcal{H}(t) = \int_0^{\infty} e^{-\lambda_k t}\left\{(-\lambda\varphi_k(a)n_2(t, a) + \varphi_k(a)\partial_t n_2(t, a))\, H\left(\frac{n_1}{n_2}(t, a)\right)\right.$$

$$\left. +\varphi_k(a)\left(\partial_t n_1(t, a) - \frac{n_1}{n_2}(t, a)\partial_t n_2(t, a)\right) H'\left(\frac{n_1}{n_2}(t, a)\right)\right\} da$$

$$= \int_0^{\infty} e^{-\lambda_k t}\left\{(-\lambda\varphi_k(a)n_2(t, a) + \varphi_k(a)(-\partial_a n_2(t, a) - B(a)n_2(t, a))H(\frac{n_1}{n_2}(t, a))\right.$$

$$\left. +\varphi_k(a)\left(-\partial_a n_1 - Bn_1 - \frac{n_1}{n_2}(-\partial_a n_2 - Bn_2)\right)(t, a)H'\left(\frac{n_1}{n_2}(t, a)\right)\right\} da$$

$$= \int_0^{\infty} e^{-\lambda_k t}\left(-\lambda\varphi_k n_2 + \partial_a \varphi_k n_2 - B\varphi_k n_2\right) H\left(\frac{n_1}{n_2}\right)(t, a)da$$

$$+ \int_0^{\infty} e^{-\lambda_k t}\varphi_k(a)\left(\partial_a n_1(t, a) - \frac{n_1}{n_2}(t, a)\partial_a n_2(t, a)\right) H'\left(\frac{n_1}{n_2}(t, a)\right) da$$

$$+e^{-\lambda_k t}\varphi_k n_2 H\left(\frac{n_1}{n_2}\right)(t, 0) + \int_0^{\infty} e^{-\lambda_k t}\varphi_k(a)\left(-\partial_a n_1 + \frac{n_1}{n_2}\partial_a n_2\right)H'\left(\frac{n_1}{n_2}\right)(t, a)da$$

$$= e^{-\lambda_k t}\left\{\int_0^{\infty} -k\varphi_k(0)B(a)n_2(t, a)H\left(\frac{n_1}{n_2}(t, a)\right) da\right.$$

$$\left. +k\varphi_k(0)H\left(\frac{n_1}{n_2}(t, 0)\right) \int_0^{\infty} B(a)n_2(t, a)da\right\}$$

$$= k\varphi_k(0)e^{-\lambda_k t}\int_0^{\infty} B(a)n_2(t, a)\left\{-H\left(\frac{n_1}{n_2}(t, a)\right) + H\left(\int_0^{\infty} B(s)\frac{n_1(t,s)}{n_2(t,0)}ds\right)\right\}da$$

$$\leq 0,$$

the last inequality following by Jensen's inequality applied to the convex function H applied to the function $f(a) = \frac{n_1}{n_2}(t, a)$, with respect to the probability measure $B(a)\frac{n_2(t,a)}{n_2(t,0)}da$.   □

Remark 3.5: In full generality, we could replace $\varphi_k(x)e^{-\lambda_k t}$ by any solution of the adjoint equation of (3.1). We can also relax the assumption on H differentiable, by taking a regularising sequence, since at the limit the terms involving H' cancel each other. For $k = 1$, we have $\varphi_k e^{-\lambda_k t} \equiv 1$: the "general relative entropy" is a standard relative entropy between n_1 and n_2.

Long-time behaviour

The long-time behaviour of the solution n may be obtained by several methods: the semi-group theory [46, 56], the use of a Laplace transform [57], the general relative entropy [34] extended recently to measure solutions [52], use of invariants [58], and very recently for measure-valued solutions Harris theorem and Doeblin's conditions [59, 60]. Whatever the method, the general fact is that under suitable assumptions, we have the convergence

$$n_k(t, a)e^{-\lambda_k t} \to N_k(a) \int_0^\infty n_k^0(a)\varphi_k(a)da$$

with λ_k and N_k defined in Proposition 3.1, and this convergence is exponentially fast in functional spaces where a spectral gap may be proved. More specifically, let us cite the following result, adapted from Ref. [55], Theorem 3.8.

Theorem 3.6: *(From Ref. [55]) Assume that there exists $a_0 > 0$, $\underline{B} > 0$, $p > 0$ and $\ell \in (p/2, p]$ such that $\forall a \in [a_0 + \mathbb{N}p, a_0\ell + \mathbb{N}p], B(a) \geq \underline{B}$. Then there exist $C > 0$, $\rho > 0$ (which can be explicitly computed) such that for any positive finite measure μ^0, there exists a unique measure-valued solution μ_t to (3.1) in a weak sense, and the following inequality holds:*

$$\|e^{-\lambda t}\mu_t - \langle \mu^0, \varphi_k\rangle N_k\|_{TV} \leq C\|\mu^0\|_{TV}e^{-\rho t},$$

where $\|\cdot\|_{TV}$ denotes the total variation norm.

Note that the assumptions are more restrictive than for the existence or the eigenproblem results, and still more restrictive to prove an exponential rate of convergence. To give an elementary intuition of this convergence, let us look at the case $B > 0$ constant: then, Proposition 3.1 shows that

$\lambda_k = (k-1)B$, $\varphi_k \equiv 1$, $N_k(a) = kBe^{-kBa}$ and a simple integration of the equation implies

$$\int_0^\infty n(t,a)da = e^{(k-1)Bt} \int_0^\infty n^0(a)da,$$

we now go back to the Duhamel formula (3.6) and find

$$n(t,a) = n^0(a-t)e^{-\int_0^t B(s+a)ds}\mathbb{1}_{\{a \geq t\}}$$
$$+ke^{-\int_0^a B(\sigma)d\sigma}\int_0^\infty B(s)n(t-a,s)ds\mathbb{1}_{\{t \geq a\}}$$

$$= n^0(a-t)e^{-Bt}\mathbb{1}_{\{a \geq t\}} + N_k(a)e^{\lambda_k t}\mathbb{1}_{\{t \geq a\}}\int_0^\infty n^0(a)\varphi_k(a)da.$$

The first term of the right-hand side vanishes exponentially fast at rate B, and the second term of the right-hand side is the expected limit.

3.2. *The renewal process*

Let us first discuss some heuristics for the convergence of empirical measures for further statistical applications. In order to extract information about $a \mapsto B(a)$, we consider the empirical distribution for a test function g defined by

$$\mathcal{E}^T(g) = |\mathcal{V}_T|^{-1} \sum_{u \in \mathcal{V}_T} g(\zeta_u^T),$$

where

$$\mathcal{V}_T = \{u \in \mathcal{U}, \ b_u \leq T, \ b_u + \zeta_u > T\}, \tag{3.8}$$

i.e. the population of cells that are alive at time T, and $\zeta_u^T = T - b_u$ is the value of the age trait at time T. We expect a law of large number as $T \to \infty$.

Heuristically, we postulate for large T the approximation

$$\mathcal{E}^T(g) \sim \frac{1}{\mathbb{E}[|\mathcal{V}_T|]} \mathbb{E}\left[\sum_{u \in \mathcal{V}_T} g(\zeta_u^T)\right].$$

Then, a classical result based on renewal theory (see Theorem 17.1, pp. 142–143 of Ref. [47]) gives the estimate

$$\mathbb{E}[|\mathcal{V}_T|] \sim \kappa_B e^{\lambda_2 T},$$

where $\lambda_2 > 0$ is the Malthusian parameter of the model, defined as the unique solution to

$$\int_0^\infty B(x)e^{-\lambda_2 x - \int_0^x B(u)du}dx = \frac{1}{2},$$

as in (3.4), and $\kappa_B > 0$ is an explicitly computable constant. As for the numerator, call χ_t the size of a particle at time t along a branch of the tree picked at random. The process $(\chi_t)_{t\geq 0}$ is Markov process with values in $[0, \infty)$ with infinitesimal generator

$$\mathcal{A}g(x) = g'(x) + B(x)\big(g(0) - g(x)\big) \tag{3.9}$$

densely defined on bounded continuous functions. It is then relatively straightforward to obtain the identity

$$\mathbb{E}\left[\sum_{u\in V_T} g(\zeta_u^T)\right] = \mathbb{E}\left[2^{N_T} g(\chi_T)\right], \tag{3.10}$$

where $N_t = \sum_{s\leq t} \mathbf{1}_{\{\chi_s - \chi_{s_-} > 0\}}$ is the counting process associated to $(\chi_t)_{t\geq 0}$. Putting together $\mathbb{E}\big[|V_T|\big] \sim \kappa_B e^{\lambda_2 T}$ and (3.10), we thus expect

$$\mathcal{E}^T(g) \sim \kappa_B^{-1} e^{-\lambda_2 T} \mathbb{E}\left[2^{N_T} g(\chi_T)\right],$$

and we anticipate that the term $e^{-\lambda_2 T}$ should somehow be compensated by the term 2^{N_T} within the expectation. To that end, following [61] (and also [32] when B is constant) one introduces an auxiliary "biased" Markov process $(\widetilde{\chi}_t)_{t\geq 0}$, with generator $\mathcal{A}_{H_B} g(x) = \mathcal{A}g(x) = g'(x) + H_B(x)\big(g(0) - g(x)\big)$ for a biasing rate $H_B(x)$ characterised by

$$f_2(x) = H_B(x) \exp\left(-\int_0^x H_B(y)dy\right),$$

with

$$f_2(x) = 2e^{-\lambda_2 x} f_1(x), \quad x \geq 0, \tag{3.11}$$

where

$$f_1(a) = B(a) \exp\left(-\int_0^a B(s)ds\right)$$

is the typical lifetime of a cell (without observation bias), or equivalently the density distribution of ζ_u, see (4.3) below. This choice (and actually this choice only) enables us to obtain

$$e^{-\lambda_2 T} \mathbb{E}\left[2^{N_T} g(\chi_T)\right] = 2^{-1} \mathbb{E}\left[g(\widetilde{\chi}_T)B(\widetilde{\chi}_T)^{-1}H_B(\widetilde{\chi}_T)\right] \tag{3.12}$$

with $\widetilde{\chi}_0 = 0$. Moreover $(\widetilde{\chi}_t)_{t \geq 0}$ is geometrically ergodic, with invariant probability $c_B \exp(-\int_0^x H_B(y)dy)dx$. We further anticipate

$$\mathbb{E}\left[g(\widetilde{\chi}_T)B(\widetilde{\chi}_T)^{-1}H_B(\widetilde{\chi}_T)\right] \sim c_B \int_0^\infty g(x)B(x)^{-1}H_B(x)e^{-\int_0^x H_B(y)dy}dx$$

$$= 2c_B \int_0^\infty g(x)e^{-\lambda_2 x}B(x)^{-1}f_1(x)dx$$

assuming everything is well-defined, by (3.11). Finally, we have $\kappa_B^{-1}c_B = 2\lambda_2$ which enables us to conclude

$$\mathcal{E}^T(g) \to \mathcal{E}(g) := 2\lambda_2 \int_0^\infty g(x)e^{-\lambda_2 x}e^{-\int_0^x B(y)dy}dx \qquad (3.13)$$

as $T \to \infty$. The convergence is in probability, with some explicit rate linked to λ_2, as given below.

Definition 3.7: The family of (real-valued) random variables $(\Upsilon_T)_{T>0}$ is asymptotically bounded in probability if

$$\limsup_{T \to \infty} \mathbb{P}(|\Upsilon_T| \geq K) \to 0 \text{ as } K \to \infty.$$

In other words, the family of distributions of the random variables Υ_T is tight, or weakly relatively compact on a neighbourhood of $T = \infty$.

Proposition 3.8: *Rate of convergence for particles living at time T (Theorem 3 in Ref. [14])* *Assume $\lambda_2 \leq 2\inf_x H_B(x)$. If B is differentiable and satisfies $B'(x) \leq B(x)^2$ and $0 < c \leq B(x) \leq 2c$ for every $x \geq 0$ and some $c > 0$, then*

$$e^{\lambda_2 T/2}\big(\mathcal{E}^T(g) - \mathcal{E}(g)\big)$$

is asymptotically bounded in probability.

3.3. The growth-fragmentation equation

When the structuring variable is the size, the so-called growth-fragmentation equation appears in many applications, from TCP-IP protocol to polymerization reactions. Let us write (2.8) here under a more general form, with a growth rate $\tau(x)$, a division rate $B(x) = \tau(x)\mathcal{B}(x)$, an a fragmentation kernel $b(y, dx)$ representing the probability distribution (in dx) of the offspring of a dividing individual of size y:

$$
\begin{cases}
\frac{\partial}{\partial t} n_k(t, dx) + \frac{\partial}{\partial x}\left(\tau(x) n_k(t, dx)\right) = -B(x)\tau(x) n_k(t, dx) \\[2mm]
\qquad\qquad + k \int_{y=x}^{\infty} B(y)\tau(y) n(t, y) b(dy, x), \qquad\qquad (3.14) \\[2mm]
\tau(0) n_k(t, 0) = 0,
\end{cases}
$$

with two conditions ensuring conservation of mass through division and division into two (this second assumption is easily relaxed but this would be meaningless in our application context):

$$
\int_0^y x b(y, dx) = \frac{y}{2}, \qquad \int_0^y b(y, dx) = 1.
$$

An important simplification often done is to restrict the equation to so-called *self-similar* division kernel b, meaning that the division place only depends on the ratio between the mother size and the daughter size: in such a case, we have

$$
b(y, dx) := \frac{1}{y} b_0\left(\frac{dx}{y}\right), \qquad \int_0^1 b_0(dz) = 1, \qquad \int_0^1 z b_0(dz) = \frac{1}{2}, \quad (3.15)
$$

and modelling considerations also lead to $b(y, dx) = b(y, y - dx)$, leading to b_0 symetric in $1/2$ (in a still more general way, we may consider a stochastic number of children, see Ref. [43]). In the study of the equation, the moments play a very important role, and the moments of order zero and one have a physical interpretation: integrating the equation, we obtain — formally at this stage

$$
\frac{d}{dt} \int n_k(t, x) dx = (k - 1) \int \tau(x) B(x) n_k(t, x) dx,
$$

which means that for $k = 1$ the number of individuals is constant — the equation is conservative — whereas for $\varepsilon = 2$ it grows due to the division process. Integrating the equation against the weight x, we have

$$
\frac{d}{dt} \int x n_k(t, x) dx = \left(\frac{k}{2} - 1\right) \int B(x)\tau(x) n_k(t, x) dx + \int \tau(x) n_k(t, x) dx,
$$

which is easily interpreted: for $k = 1$ the mass increases with growth but decreases with division, whereas for $k = 2$ it only increases with growth. More generally, moments of order p show a balance between growth and division:

$$
\frac{d}{dt} \int_0^{\infty} x^p n dx = \int_0^{\infty} \left(p - B(x) x \left(1 - k \int_0^x \frac{y^p}{x^p} b(x, dy)\right)\right) x^{p-1} \tau(x) n dx.
$$

This balance between growth and division leads to the main asymptotic behaviour to be expected: the convergence to a steady size-distribution profile, with exponential growth in time for the population case $k = 2$, at an exponential rate of convergence if a spectral gap is proved. This study begins in the 1980's with the work by Diekmann, Heijmans, Thieme and Gyllenberg and Webb [62,63], based on the theory of semigroups, and generally carried out under the assumption of a compact support for the size so that general theorems may apply in a more direct way. This has been then generalised by several authors, see Refs. [55,56]. Explicit solutions, for power law rates and specific fragmentation kernels have also been studied [64–66].

Eigenelements

The eigenvalue problem and its adjoint are as follows:

$$
\begin{cases}
\frac{\partial}{\partial x}((\tau N_k)(x)) + \lambda_k N_k(x) = -(\tau B N_k)(x) + k \int_x^\infty (\tau B N_k)(y)b(y,x)dy, \\[2mm]
\tau N_k(x = 0) = 0, \qquad N_k(x) \geq 0, \qquad \int_0^\infty N_k(x)dx = 1, \\[2mm]
-\tau(x)\frac{\partial}{\partial x}(\varphi_k(x)) + \lambda_k \varphi_k(x) = B(x)\tau(x)(-\varphi_k(x) + k \int_0^x b(x,y)\varphi_k(y)dy), \\[2mm]
\varphi_k(x) \geq 0, \qquad \int_0^\infty \varphi_k(x)N_k(x)dx = 1.
\end{cases}
$$

$$(3.16)$$

These equations are meant in a weak sense, *i.e.* we look for $N_k \in L^1((0,\infty),dx)$ such that $\forall \phi \in \mathcal{C}_c^\infty([0,\infty))$ we have

$$
-\int_0^\infty \tau N_k \partial_x \phi dx + \lambda \int_0^\infty \phi dx = \int_0^\infty B\tau N_k \left(k \int_0^x \phi(y)b(x,dy) - \phi(x)\right) dx,
$$

and $\varphi \in W_{loc}^{1\infty}$ is solution almost everywhere. For existence and uniqueness of eigenelements, the following assumptions are among the most general ones — except the fact that rates are at most polynomially growing, which is relaxed in Theorem 3.11 below, but which was useful in the proof of [36] based on moment estimates:

$$Supp(b(y,\cdot)) \subset [0,y], \qquad \int_0^y b(y,dx) = 1,$$

$$\int_0^y xb(y,dx) = \frac{y}{2}, \qquad \int_0^y \frac{x^2}{y^2}b(y,dx) \le c < \frac{1}{2},$$

$$\tau, B \in \mathcal{P} := \left\{ f \ge 0 : \exists\, \mu, \nu \ge 0, \limsup_{x\to\infty} x^{-\mu} f(x) < \infty, \liminf_{x\to\infty} x^\nu f(x) > 0 \right\},$$

$$\tau B \in L^1_{loc}((0,\infty)), \quad \exists\, \alpha_0 \ge 0, \quad \tau \in L^\infty_{loc}([0,\infty), x^{\alpha_0} dx),$$

$$\forall K \text{ compact on } (0,\infty), \quad \exists\, m_K > 0, \quad \tau(x) \ge m_K \text{ a.e. } x \in K,$$

$$\exists\, B_0 \ge 0, C > 0, \gamma \ge 0, \ Supp(B) = [b,\infty), \int_0^x b(y,dz) \le \min\left(1, C(\tfrac{x}{y})^\gamma\right),$$

$$B, \frac{x^\gamma}{\tau} \in L^1_0 := L^1_{loc}([0,\cdot)), \qquad \lim_{x\to\infty} xB = \infty.$$

$$\tag{3.17}$$

Theorem 3.9: *(From Ref. [36]) Under the balance assumptions (3.17) on τ, B and b, there exists a unique triplet $(\lambda_k, N_k, \varphi_k)$ with $\lambda_2 > 0$, $\lambda_1 = 0$, $\varphi_1 \equiv 1$ solution of the eigenproblem (3.16) and*

$$x^\alpha \tau N_k \in L^p(\mathbb{R}^+), \quad \forall \alpha \ge -\gamma, \quad \forall p \in [1,\infty], \quad x^\alpha \tau N_k \in W^{1,1}(\mathbb{R}^+),$$

$$\exists\, p > 0 \text{ s.t. } \frac{\varphi_2}{1+x^p} \in L^\infty(\mathbb{R}^+), \quad \tau \frac{\partial}{\partial x}\varphi_2 \in L^\infty_{loc}(\mathbb{R}^+).$$

We notice that the assumptions on B at ∞ are very similar to the ones for the renewal equation. If $B(x) = x^\gamma$, the assumptions on B are satisfied for $1 + \gamma > 0$, see Ref. [67]. The proof is based on standard theorems for regularized equations (Krein-Rutman or Perron-Frobenius), the compactness obtained by successive moments estimates, and uniqueness by using general relative entropy inequalities.

Finally, a quick computation shows that in the case of exponential growth $\tau(x) \equiv x$, and of division into two equally-sized daughters $b(y,x) = \delta_{x=\frac{y}{2}}$ we have $\varphi_2(x) \equiv Cx$ with $C > 0$ a normalisation constant, and $N_1(x) = C'x N_2(x)$ with $C' > 0$ another normalisation constant.

General Relative Entropy

Lemma 3.10: *(General Relative Entropy Inequality [35], Theorem 2.1) Let n_1, n_2 be two solutions of (3.14), with $n_2(t,x) > 0$ for all* *time and size, $H : \mathbb{R} \to [0,\infty)$ positive, differentiable and convex, and $\int_0^\infty \varphi(x) n_2(0,x) H\left(\frac{n_1}{n_2}(0,x)\right) dx < \infty$. Then we have*

$$\mathcal{H}(t) := \int_0^\infty \varphi(x) e^{-\lambda t} n_2(t,x) H\left(\frac{n_1}{n_2}(t,x)\right) dx, \quad \frac{d\mathcal{H}}{dt} = -D^H[n_1, n_2] \le 0,$$

with

$$D^H[n_1, n_2](t) = \int_0^\infty \int_0^\infty k\varphi(x) e^{-\lambda t} n_2(t,y) \tau(y) B(y) b(y,x)$$

$$\left\{ H\left(\frac{n_1}{n_2}(t,y)\right) - H\left(\frac{n_1}{n_2}(t,x)\right) - H'\left(\frac{n_1}{n_2}(t,x)\right) \left\{ \frac{n_1}{n_2}(t,y) - \frac{n_1}{n_2}(t,x) \right\} \right\} dx dy.$$

We let the reader check directly this computation or refer to Ref. [68] or [34]. We also observe that the entropy for the renewal equation may be viewed as a specific case of this inequality: it suffices to take $b(y,x) = \delta_{x=0}$ and $\tau(x) = 1$. As for the renewal equation, it is possible to prove long-term convergence by means of this entropy inequality [34, 68], and exponential rate of convergence through entropy-entropy dissipation inequalities [69], *i.e.* if we can bound $-D^H$ by a quantity depending on $\mathcal{H}$.

Long time asymptotics: the central case of asynchronous exponential growth

Many methods have been developed to study the long time asymptotics of growth-fragmentation equations, from semi-group theory, general relative entropy, to methods inspired by stochastic processes such as several very recent studies [70, 71]. Let us cite here a recent result, carried out only for the two extreme cases of uniform (*i.e.* $b_0 = 2\mathbb{1}_{[0,1]}$) or equal mitosis (*i.e.* $b_0 = 2\delta_{1/2}$) fragmentation kernels, but relatively general for the assumption on the fragmentation rate, which may grow faster than polynomially.

Assumption 1: (**Assumptions for a spectral gap, see Ref. [71]**)

- $b(y,x) = \frac{1}{y}\mathbb{1}_{\{x \le y\}}$ or $b(y,x) = \delta_{x=\frac{y}{2}}$.
- τ is locally Lipschitz, $g(x) = O(x)$ around ∞, $g(x) = O(x^{-\xi})$ around 0 with $\xi \ge 0$,
- $B : (0,\infty) \to [0,\infty)$ is continuous,
- $\int_0 B dx < \infty$, $\quad \lim_{x\to 0} xB(x) = 0$, $\quad \lim_{x\to\infty} xB(x) = \infty$.
- If $b(y,x) = \delta_{x=\frac{y}{2}}$, the growth rate τ must moreover satisfy
  $- \omega g(x) < g(\omega x)$ for all $x > 0$ and $0 < \omega < 1$,

- $H(z) := \int_0^z \tau^{-1}(z)dz < \infty$ for all $z > 0$,
- $\lim\limits_{z\to\infty} H^{-1}(z+r)/H^{-1}(z) = 1$.

Restricted to $\tau(x) = x^\nu$, we see that the assumptions imply $\nu \leq 1$ for the uniform kernel, $\nu < 1$ for the equal mitosis kernel, and allow any growth for B at infinity: compared to the assumptions for the existence of eigenelements, the main restriction, apart from the specific shapes of the fragmentation kernel, is that we cannot consider superlinear growth rates, since then the cell sizes may explode in finite time.

Theorem 3.11: (Theorem 1.3 from Ref. [71]) *Under Assumptions 1, there exists a unique eigentriplet $(\lambda_2, N_2, \varphi_2)$ solution to (3.16). Let us denote, for $k \leq 0$ and $K > 1$, the following weighted total variation norm*

$$\|\mu\|_{k,K} := \int_0^\infty (x^k + x^K)|\mu|(dx).$$

Then for n_0 a non-negative finite measure satisfying $\|n_0\|_{k,K} < \infty$, there exists a unique measure-valued solution n_2 to (3.14) and it satisfies, for some $C > 0$ and $\rho > 0$,

$$\|e^{-\lambda_2 t}n_2(t,\cdot) - <\varphi_2, n_0 > N_2\|_{k,K} \leq Ce^{-\rho t}\|n_0 - <\varphi_2, n_0 > N_2\|_{k,K},$$

with the following possible choices for k and K:

1. *if $\int_0 \tau(x)^{-1}dx < \infty$, take $k = 0$ and any $K > 1 + \xi$,*
2. *if $\int_0 \tau(x)^{-1}dx = \infty$, take any $k \in (-1,1)$ and $K > 1 + \xi$,*
3. *if $\tau(x) = x$, take any $k \in (-1,1)$ and $K > 1$.*

The assumptions on the state space where the convergence holds are crucial to obtain the exponential speed of convergence, which is linked to a spectral gap. Specifically, P. Michel, S. Mischler and B. Perthame proved convergence — without speed — in the weighted space $L^1(\varphi_2 dx)$ [35], which is the most natural space to prove convergence results through the general relative entropy inequality; but under the assumption of bounded fragmentation rates, E. Bernard and P. Gabriel proved that there exists no spectral gap in this space: the convergence may hold arbitrarily slowly for well-chosen initial conditions, see Theorems 1.2 and 4.1 in Ref. [72]. Among other important results that we cannot review in detail here, let us cite fine estimates on the eigenvector and adjoint eigenvector [69, 73], semigroup approaches [56], probabilistic approaches [74, 75].

Long time asymptotics: other cases

When the balance assumptions between growth and division around zero and around infinity fail to be satisfied, other types of asymptotic behaviour may happen, leading to mass escape towards zero (dust formation or shattering) or infinity (gelation). Let us focus on the case of exponential growth $\tau(x) = x$, interesting in several ways: as already said, it is the idealised growth rate for many unicellular organisms, like bacteria; it is also the limit case before characteristic curves grow to infinity in finite time; last but not least, it appears by a change of variables when studying the asymptotic trend to a self-similar profile for the pure fragmentation equation, see Ref. [13] Theorem 3.2. Assuming a power law for the division rate $B(x) = x^\gamma$, we can classify the anomalous asymptotic behaviours according to the value of γ.

- $\gamma < -1$: in such a case, there is a loss of mass by dust formation in finite time called *shattering* [44, 76–80], non-uniqueness of solutions [81, 82].
- $\gamma = -1$: this is a case where the *time-dependent* division rate $\beta(x) = B(x)\tau(x)$ is constant. It is a limit case, where there is neither loss of mass in finite time nor convergence to a steady profile and exponential growth, since each moment of the equation grows or decays exponentially with a specific rate, see Ref. [83]. This is interesting from a modelling perspective because it explains the fact that a model with both exponential growth in size and size-independent division (for instance an age-structured division rate for exponentially growing cells) is irrelevant, leading to a non-realistic exponential behaviour since no steady profile in size may be obtained [19]. Generalisations of this limit case to $\tau(x) = x^{1+\gamma}$ and $B(x)\tau(x) = x^\gamma$ with $\gamma > 0$, leading to blow-up in finite time, has been done by M. Escobedo [64, 84] and J. Bertoin and A. Watson for the corresponding stochastic processes [85].
- $\gamma > -1$: in general, convergence theorems such as Theorem 3.11 are valid, however for very specific division kernel such as the idealised equal mitosis case, the solution may converge to a cyclic behaviour. In such a case, we still have a general relative entropy inequality given by Lemma 3.10, but a simple computation shows that the entropy dissipation D^H vanishes not only for $\frac{n_1}{n_2}$ constant, but also for any ratio satisfying

$$\frac{n_1(x)}{n_2(x)} = \frac{n_1(2x)}{n_2(2x)},$$

which is a kind of periodicity condition. The best intuition on what happens here comes from the underlying stochastic branching tree: all descendants of a given cell of size x_0 at time 0 live on the countable set of curves $x_0 e^t 2^{-n}$, due to the very specific relation between growth and division, whereas for other growth or division the times of division account, leading to a kind of dissipativity. This case has been studied by semigroup theory for compact support in size by G. Greiner and R. Nagel [86], and extended and revisited in Ref. [87] where the following explicit asymptotic result has been proved — see also Ref. [88] for extension to measure solutions.

Theorem 3.12: *(Theorem 2.3 in Ref. [87]) Let $\tau > 0$ and define $\tau(x) = x$, $b(y, x) = 2\delta_{x=\frac{y}{2}}$, and B such that*

$$\begin{cases} B : (0, \infty) \to (0, \infty) \text{ is measurable, } B \in L^1_{loc}([0, \infty)), \\ \exists \gamma_0, \gamma_1, K_0, K_1, x_0 > 0, \quad K_0 x^{\gamma_0} \leq xB(x \geq x_0) \leq K_1 x^{\gamma_1}. \end{cases} \tag{3.18}$$

Theorem 3.9 holds, but there also exists a countable set of non-positive dominant eigentriplets defined, for $m \in \mathbb{Z}$, by

$$\lambda_k^m = (k-1) + \frac{2im\pi}{\ln 2}, \quad N_k^m(x) = x^{-\frac{2im\pi}{\ln 2}} N_k^m(x), \quad \varphi_k^m(x) = c_m x^{k-1+\frac{2im\pi}{\ln 2}},$$

with c_m normalisation constants. All the quantities $< n_k(t, \cdot), \varphi_k^m e^{-\lambda_k^m t} >$ are then conserved, and for any $n_0 \in L^2([0, \infty), x^{k-1}/N_k^0(x)dx)$, the unique solution $n_k(t, x) \in C([0, \infty), L^2([0, \infty), x^{k-1}/N_k^0(x)dx))$ to (3.14) satisfies

$$\int_0^\infty \left| n_k(t, x) e^{-\kappa(k-1)t} - \sum_{m=-\infty}^\infty < n_0, \varphi_k^m > N_k^m(x) e^{\frac{2im\pi}{\ln 2}t} \right|^2 \frac{x \, dx}{N_k^0(x)} \xrightarrow{t \to \infty} 0.$$

A numerical scheme needs to be non-dissipative to capture the oscillations, for instance by splitting transport and fragmentation and by using a geometric grid, see Sec. 3 in Ref. [87]; another way would be to use a splitting particle method [89].

3.4. *Structured population equations and processes*

For more general models and methods, several excellent books [34, 46, 90] have been written together with an extensive literature, mainly for linear but also for nonlinear [54, 91] cases, from a PDE or a stochastic point of view. Let us mention here only the expected asymptotic behaviour, given by the eigenvector(s) linked to the dominant eigenvalue(s), for the models cited in Sec. 2.3.

The incremental/adder model

The eigenvalue problem linked to the system Eqs. (2.10) and (2.11) may be written as follows:

$$\begin{cases} \lambda_k N_k + \frac{\partial}{\partial z}(\kappa x N_k) + \frac{\partial}{\partial x}(\kappa x N_k) = -\kappa x B(z) N_k(z, x), \\ \kappa x N_k(0, x) = 4k\kappa x \int_0^\infty B(z) N_k(z, 2x) dz, \end{cases}$$

(3.19)

with $\lambda_1 = 0$ as usual, and $\lambda_2 = \kappa$ as for the growth-fragmentation equation, due to the linear growth rate. Existence and uniqueness of a dominant positive eigenvalue and eigenvector has been recently studied in Ref. [92], for more general fragmentation kernels than just the diagonal kernel. Note also that in this specific case, we also have a countable set of dominant (not positive) eigenvalues, so that an equivalent of Theorem 3.12 may be obtained.

Generalisations

Several types of generalisations have been studied, for instance with varying growth rates [37], or with a maturity variable added to the renewal equation [54], size and age structured models [93, 94] etc. Let us only write here the eigenvalue problem related to Eqs. (2.12) and (2.13) — whose study once more lies beyond the scope of this chapter:

$$\lambda_k N_k + \frac{\partial}{\partial z}(\tau_z(z, x) N_k) + \frac{\partial}{\partial x}(\tau_x(x, \kappa) N_k) = -\beta(z, x, \kappa) N_k(z, x, \kappa),$$

(3.20)

$$\tau_x N_k(z, 0, \kappa) = 0, \qquad N_k \geq 0, \qquad \iiint N_k \, dz dx d\kappa = 1,$$

$$\tau_z(x, \kappa) N_k(0, x, \kappa) =$$

(3.21)

$$2k \int_0^\infty \int_0^\infty \int_0^\infty \theta(\kappa', \kappa) b(y, x) \beta(z, y, \kappa) N_k(z, y, \kappa') dz dy d\kappa'.$$

4. Model calibration: statistical estimation of the division rate

In Sec. 1.3, we have noticed that, contrarily to the growth rate or even to the division kernel, the division rate cannot be inferred from direct measurements, even from individual dynamics data. We thus face a typical inverse problem: How to estimate the division rate B from data on a population, which follows — we assume — the dynamics given by one of the structured population model described above?

A first and major idea [12] consists in taking advantage of the asymptotic analysis carried out in Sec. 3: we consider that at any time of the experiment, the population has already reached its steady asymptotic regime, *i.e.* for the observation scheme k that the population is aligned along the dominant eigenvector N_k of the model under consideration: age, size, increment, or more general model. This is well-justified by the theoretical analysis above: the trend being exponentially fast under fairly general assumptions, the not-asymptotic regime concerns in most experimental cases only a negligible part of the data collected.

We recall (see Sec. 1.3) that there are two types of datasets, each being related to a different inverse problem:

- **Individual dynamics data collection**: following the trajectory of each individual allows us to measure the dividing and newborn cells. Intuitively, one feels that this allows a relatively direct estimation of the division rate. This type of data may concern genealogical (through *e.g.* microfluidic device) as well as population (microcolony growth) observation. A difficulty in the population dynamics observation is the selection bias [14].
- **Population point data**: we can observe, at given timepoints, samples of some structuring variables such as size (or age, fluorescent label [95] or whatsoever), which are then related to the empirical distribution (2.2), itself related to $f_t(\cdot) = \frac{n_k(t,\cdot)}{\int n_k(t,\cdot)d\cdot}$ thanks to a representation like in Proposition 2.2. We may moreover assume an approximation of the form $f_t \approx N_k$ by the use of a time-asymptotic result such as Theorem 3.6 or 3.11. Alternatively, we can keep up with a stochastic approach, linking directly the stochastic measure to its limit through probabilistic results such as [14, 37, 75]. We then address the inverse problem consisting in estimating B from measurements of N_k; one feels immediately that such an approach requires more analysis and, since the available information is less rich, that the inverse problem is more ill-posed.

Note that even in the case of individual dynamics collection, it may be more interesting to use the second approach: if the data are more numerous or less noisy, this may compensate the fact that the information they contain is poorer. In our example of *E. coli*, both approaches are possible, which allows to compare their accuracy in practice.

4.1. *Estimating an age-dependent division rate*

As for the previous Secs. 2 and 3.1, the age-structured model is somehow the simplest model along our line of models for which explicit computations can be conducted. We review here the different types of inverse problems we have to solve, depending on the type of data available; we will find all the same problems for the other models.

4.1.1. *Individual dynamics data*

Individual dynamics data, stochastic viewpoint. Let us assume that we have data such as shown in Fig. 1 or 2: at short time intervals, we observe the age of cells, so that we observe

$$\{\zeta_u, \quad u \in \mathcal{U}_k\},$$

for some subtree $\mathcal{U}_k \subset \mathcal{U}$, with $\mathcal{U}$ being defined in (2.1). For the genealogical observation ($k = 1$), we define

$$\mathcal{U}_1 = \{u_\ell \in \{0,1\}^\ell, \quad u_{\ell+1} = (u_\ell, u^+), \quad 0 \le \ell \le n \quad u^+ \in \{0,1\}\}, \tag{4.1}$$

with $u^+ \in \{0,1\}$ chosen uniformly at random, so that $u_{\ell+1}$ is offspring of u_ℓ, and n is a fixed number given by the experimentalist. In practice, we gather several such trees. For the population observation ($k = 2$), the trees are defined by a final time $T > 0$ fixed by the experimentalist, so that we observe

$$\mathcal{U}_2 = \{u \in \mathcal{U}, b_u + \zeta_u \le T\}, \tag{4.2}$$

where b_u is the birth time of the cell: we observe all the lifetimes of cells which have divided before T. We see that the number of cells is stochastic for this second case, and there is a selection bias: we will observe more descendants of cells which have divided quickly, see Sec. 3.

Individual dynamics data, stochastic viewpoint, genealogical observation. This case is relatively straightforward: as already observed, since

$$\mathbb{P}(\zeta_u \in [a, a + da] \,|\, \zeta_u \ge a) = B(a)da,$$

we obtain that the probability distribution of a lifetime ζ_u is given by

$$\mathbb{P}(\zeta_u \in da) = f_1(a)da = B(a)\exp\left(-\int_0^a B(s)ds\right)da. \tag{4.3}$$

In the case of a genealogical observation, the subtree is deterministic: there is no selection bias and the lifetime of each cell is independent from the others: we observe a sample of n cells having divided at ages which are the realizations of $\{\zeta_u, u \in \mathcal{U}_1\}$, as independent random variables with common density f. Moreover, as soon as $\int^\infty B = \infty$, we have the survival analysis representation

$$B(a) = \frac{f_1(a)}{\int_a^\infty f_1(s)ds} = \frac{f_1(a)}{S_1(a)}, \tag{4.4}$$

where S_1 is the survival function, as a simple inversion of the formula (4.3) given above. In other application contexts, B is called a hazard function. In this simple formula, we notice here three important facts, that will be found throughout our study:

- estimating B has the same complexity as estimating the density f_1. This stems from the fact that the survival function $S_1(a)$ can be estimated at rate $\sqrt{n}$ by its empirical counterpart, hence only the numerator in the right-hand side of (4.4) is a genuine nonparametric estimation problem.
- The estimation of $B(a)$ becomes harder as a increases, since $S(a)$ vanishes when a tends to infinity.
- We can also interpret our observations directly on the eigenvector equation N_1 : the proportion of dividing cells being $B(a)N_1(a)$, we find B by writing simply

$$B(a) = \frac{B(a)N_1(a)}{N_1(a)},$$

and using the equation again we find that $N_1(a) = Ce^{-\int_0^a B(s)ds} = CS_1(a)$, with $C > 0$ a normalisation constant, so that we are back to (4.4).

We make the first point rigorous by recalling a standard statistical estimation result. Let $K : [0, \infty) \to \mathbb{R}$ denote a well-located kernel of order $\ell \geq 0$, namely

$$K \in \mathcal{C}_c^0(\mathbb{R}), \qquad \int_0^\infty a^k K(a)da = \mathbb{1}_{\{k=0\}} \text{ for } k = 0, \dots, \ell. \tag{4.5}$$

The existence of such an oscillating kernel for arbitrary ℓ is standard, see *e.g.* the textbook [96]. For $h > 0$ the bandwidth, define $K_h(a) = h^{-1}K(h^{-1}a)$ and

$$\widehat{B}_{n,h}(a) = \frac{\sum_{u \in \mathcal{U}_1} K_h(a - \zeta_u)}{\sum_{u \in \mathcal{U}_1} \mathbb{1}_{\{\zeta_u \geq a\}}}, \tag{4.6}$$

(and set 0 if none of the ζ_u are above a). The *bias* of f_1 at a relative to the approximation kernel K is defined as

$$\mathfrak{b}_h(f_1)(a) = \left| \int_{[0,\infty)} f_1(a') K_h(a - a') da' - f_1(a) \right|.$$

Proposition 4.1: *We have*

$$\mathbb{E}\left[\left| |\mathcal{U}_1|^{-1} \sum_{u \in \mathcal{U}_1} K_h(a - \zeta_u) - f_1(a) \right|^2 \right]$$

$$\leq \mathfrak{b}_h(f_1)(a)^2 + (nh)^{-1} \sup_{a - a' \in \mathrm{Supp}(K)} B(a') \int_{[0,\infty)} K(a')^2 da' \qquad (4.7)$$

and

$$\mathbb{E}\left[\left| |\mathcal{U}_1|^{-1} \sum_{u \in \mathcal{U}_1} \mathbb{1}_{\{\zeta_u \geq a\}} - S_1(a) \right|^2 \right] \leq \tfrac{1}{4} n^{-1}. \qquad (4.8)$$

Proof: The first part is obtained by noticing that $\int_{[0,\infty)} f_1(a') K_h(a - a') da' = \mathbb{E}[K_h(a - \zeta_u)]$, and using that the variables $K_h(a - \zeta_u) - \mathbb{E}[K_h(a - \zeta_u)]$ are independent and identically distributed, with common variance bounded above by $h^{-1} \int_{[0,\infty)} K_h(a - a')^2 f_1(a') da' \leq \sup_{a - a' \in \mathrm{Supp}(K)} B(a') \int_{[0,\infty)} K(a')^2 da'$. The result is simply a combination of this observation and the act that the variance of the sum of independent random variables is the sum of its variances. The second part easily follows, noticing now that $\mathbb{1}_{\{\zeta_u \geq a\}}$ is a Bernoulli random variable with expectation $S_1(a)$ and variance (always) bounded by $1/4$. $\qquad \square$

Assuming that B has smoothness of order $s > 0$ around a, in a Hölder sense for instance, we then have $\mathfrak{b}_h(f_1)(a) \lesssim h^s$ as soon as $\ell \geq s - 1$, and therefore the estimator $|\mathcal{U}_1|^{-1} \sum_{u \in \mathcal{U}_1} K_h(a - \zeta_u)$ has pointwise squared risk of order $h^{2s} + (nh)^{-1}$ in the following sense:

$$\left(\inf_{h > 0} (h^s + (nh)^{-1/2}) \right)^{-1} \left(\widehat{B}_{n,h}(a) - B(a) \right)$$

is bounded in probability as $n \to \infty$.

Finding the optimal bandwidth h leads to the classical rate $\inf_{h > 0}(h^s + (nh)^{-1/2}) \approx n^{-s/(2s+1)}$ in nonparametric estimation, which is always a slower rate of convergence than $n^{-1/2}$. Combining the two estimates (4.7) and (4.8) for an optimal bandwidth, we see that $\widehat{B}_{n,h}(a)$ estimates $B(a)$ with optimal (normalised) order $n^{-s/(2s+1)}$. These are classical results in nonparametric estimation, see *e.g.* Refs. [96, 97] and the references therein, in particular regarding data driven choices of h, since the smoothness $s > 0$ is only a mathematical construct that has no real meaning in practice.

Individual dynamics data, stochastic viewpoint, population observation. Following in spirit Sec. 3.2 but now with data extracted from $\mathcal{U}_2$, we first look for the behaviour of empirical sums of the form

$$\mathcal{E}^T(g, \mathcal{U}_2) = \frac{1}{|\mathcal{U}_2|} \sum_{u \in \mathcal{U}_2} g(\zeta_u),$$

for nice (say bounded) test functions $g : [0, \infty) \to \mathbb{R}$. As in Sec. 3.2, we also have a many-to-one formula that now reads

$$\mathbb{E}\left[\sum_{u \in \mathcal{U}_2} g(\zeta_u^T)\right] = \mathbb{E}\left[\sum_{u \in \mathcal{U}_2} g(\zeta_u)\right] = 2^{-1} \int_0^T e^{\lambda_2 s} \mathbb{E}\left[g(\widetilde{\chi}_s) H_B(\widetilde{\chi}_s)\right] ds,$$

(4.9)

where $(\widetilde{\chi}_t)_{t \geq 0}$ is the auxiliary one-dimensional auxiliary Markov process with generator $\mathcal{A}_{H_B}$, see (3.9), where H_B is characterised by (3.11) above. Assuming again ergodicity, we approximate the right-hand side of (4.9) and obtain (at least heuristically)

$$\mathbb{E}\left[\sum_{u \in \mathcal{U}_2} g(\zeta_u)\right] \sim c_B 2^{-1} \frac{e^{\lambda_2 T}}{\lambda_2} \int_0^\infty g(x) H_B(x) e^{-\int_0^x H_B(u) du} dx$$

$$= c_B \frac{e^{\lambda_2 T}}{\lambda_2} \int_0^\infty g(x) e^{-\lambda_2 x} f_1(x) dx,$$

since $H_B(x) \exp(-\int_0^x H_B(y) dy) = 2e^{-\lambda_2 x} f_1(x)$ by (3.11). We again have an approximation of the type $\mathbb{E}[|\mathcal{U}_2|] \sim \kappa_B e^{\lambda_2 T}$ with another constant κ_B' and we eventually expect

$$\mathcal{E}^T(g, \mathcal{U}_2) \sim \mathring{\mathcal{E}}(g) := \frac{c_B}{\lambda_2 \kappa_B'} \int_0^\infty g(x) e^{-\lambda_2 x} f_1(x) dx = 2 \int_0^\infty g(x) e^{-\lambda_2 x} f_1(x) dx$$

as $T \to \infty$, where the last equality stems from the identity $c_B = 2\lambda_2 \kappa_B'$ that can be readily derived by picking $g = 1$ and using (3.11) together with the fact that f_2 is a density function.

Proposition 4.2: *(Rate of convergence for particles living at time T — Theorem 4 in Ref. [14])* Assume $\lambda_2 \leq 2 \inf_x H_B(x)$, and B differentiable satisfying $B'(x) \leq B(x)^2$ and $0 < c \leq B(x) \leq 2c$ for every $x \geq 0$

and some $c > 0$. Then

$$e^{\lambda_2 T/2}\big(\mathcal{E}^T(g) - \mathring{\mathcal{E}}(g)\big)$$

is asymptotically bounded in probability.

Estimation: Step 1). Reconstruction formula for $B(a)$. We have

$$B(a) = \frac{f_1(a)}{1 - \int_0^a f_1(y)dy} = \frac{2^{-1}f_2(a)e^{\lambda_2 a}}{1 - 2^{-1}\int_0^a f_2(y)e^{\lambda_2 y}dy} \qquad (4.10)$$

and from the definition $\mathring{\mathcal{E}}(g) = 2\int_0^\infty g(x)e^{-\lambda_2 x}f_1(x)dx =$ we obtain the formal reconstruction formula

$$B(a) = \frac{\mathring{\mathcal{E}}\big(2^{-1}e^{\lambda_2 \cdot}\delta_a(\cdot)\big)}{1 - \mathring{\mathcal{E}}\big(2^{-1}e^{\lambda_2 \cdot}1_{\{\cdot \leq a\}}\big)} \qquad (4.11)$$

where $\delta_a(\cdot)$ denotes the Dirac function at x. Therefore, taking g as a weak approximation of δ_a via a kernel, we obtain a strategy for estimating $B(a)$ replacing $\mathring{\mathcal{E}}(\cdot)$ by its empirical version $\mathcal{E}^T(\mathcal{U}_2, \cdot)$.

Estimation: Step 2). Construction of a kernel estimator and function spaces. Let $K : [0, \infty) \to \mathbb{R}$ be a kernel function. For $h > 0$, set $K_h(x) = h^{-1}K(h^{-1}x)$. In view of (4.11), we define the estimator

$$\widehat{B}_{T,h}(a) = \frac{\mathcal{E}^T\big(\mathcal{U}_2, 2^{-1}e^{\lambda_2 \cdot}K_h(a - \cdot)\big)}{1 - \mathcal{E}^T\big(\mathcal{U}_2, 2^{-1}e^{\lambda_2 \cdot}1_{\{\cdot \leq a\}}\big)} \qquad (4.12)$$

on the set $\mathcal{E}^T\big(\mathcal{U}_2, 2^{-1}e^{\lambda_2 \cdot}1_{\{\cdot \leq a\}}\big) \neq 1$ and 0 otherwise. Thus $\widehat{B}_{T,h}(a)$ is specified by the choice of the kernel K and the bandwidth $h > 0$.

Performances of the Estimator. We are ready to give the rate of convergence of $\widehat{B}_T(a)$ for a restricted to a compact interval $\mathcal{D}$, uniformly over Hölder balls $\mathcal{H}_{\mathcal{D}}^s$.

Proposition 4.3: *(Upper rate of convergence, Theorem 7 in Ref. [14])* In the same setting as in Proposition 4.2, specify $\widehat{B}_{T,h}$ with a kernel satisfying (4.5) for some $\ell > 1$ and

$$h = \widehat{h}_T = \exp\big(-\tfrac{1}{2s+1}\lambda_2 T\big)$$

for some $s \in (1, \ell + 1)$. Then

$$e^{\lambda_2 \frac{s}{2s+1}T}\big(\widehat{B}_{T,h}(a) - B(a)\big)$$

is asymptotically bounded in probability if B is s-Hölder in a neigbourhood of a.

This rate is indeed optimal in a minimax sense, see Theorem 8 in Ref. [14], where the problem of estimating λ_2 is also considered. The proof of Proposition 4.3 is detailed in Ref. [14]. In a more condensed way, they can also be found in Ref. [98].

Individual dynamics data, deterministic viewpoint. In the field of "deterministic" inverse problems, we model the noise by assuming that we observe data in a certain metric space up to an error ε according to this metric. In our case, this means that we first assume that the population has reached its steady asymptotic behaviour given by Theorem 3.6, and second that there exists a (known) noise level $\varepsilon > 0$, and that we observe the distribution of ages of dividing cells, defined by

$$f_k(a) := \frac{B(a)N_k(a)}{\int B(a)N_k(a)da},$$

up to a noise, *i.e.* the measurement $H_k^\varepsilon(a)$ is such that

$$\|f_k^\varepsilon - f_k\|_{W^{-s,p}([0,\infty))} \leq \varepsilon,$$

with $s \geq 0$, $1 \leq p \leq \infty$ and $W^{-s,p}([0,\infty))$ the corresponding Sobolev space. Integrating (3.3) to express $N_k(a)$ in terms of $B(a)N_k(a)$ and λ_k, chosen with $k = 1$ or $k = 2$ according to the observation scheme considered, we find the formula

$$B(a) = \frac{B(a)N_k(a)}{N_k(a)} = \frac{B(a)N_k(a)}{e^{-\lambda_k a}\int_a^\infty B(s)N_k(s)e^{\lambda_k s}ds} = \frac{f_k(a)}{e^{-\lambda_k a}\int_a^\infty f_k(s)e^{\lambda_k s}ds},$$
$$(4.13)$$

where we recognize (4.4) if $k = 1$ and (4.11) if $k = 2$. This naturally leads us to define an estimate B_ε by replacing in this formula f_k by f_k^ε, and add a threshold condition for the denominator, as done above by considering compact intervals. If $s = 0$, *i.e.* if the noise lies in $L^p([0,\infty))$, we do not need to regularize this estimate: the problem is well-posed, and B_ε provides directly an estimate for B in a space $L^p([0,\infty))$ weighted by N_k. If either $s < 0$ or we want an estimate for B in some $W^{m,p}$ space with $m > 0$, then a regularization is needed: in exactly the same spirit as for kernel density estimation, we can define

$$f_k^{\varepsilon,h} = K_h * f_k^\varepsilon, \qquad (4.14)$$

and we have the following result, deterministic version of the above Propositions 4.1 and 4.2.

Proposition 4.4: *Let K a kernel satisfying (4.5), $K \in C_b^1(\mathbb{R})$, and $K_h(\cdot) = 1/hK(\cdot/h)$. Let $1 \le p \le \infty$ and $\theta \in [0,1]$. We have the estimate*

$$\|f_k^{\varepsilon,h} - f_k\|_{L^p} \le \|K_h * f_k - f_k\|_{L^p} + C(K)h^{-\theta}\|f_k - f_k^\varepsilon\|_{W^{-\theta,p}},$$

where $C(k)$ is a constant depending only on the kernel K and on its derivative.

We do not specify here the standard machinery to obtain an estimate for B from the estimate for f_k : it consists in dividing $f_k^{\varepsilon,h}$ by N_k^ε — we do not need any regularisation for the denominator, thanks to the integral and to the choice $\theta \le 1$ — and then thresholding. See for instance Ref. [99].

We then find that, for a noise ε in the space $W^{-\theta,p}$, *i.e.* if we have

$$\|f_k - f_k^\varepsilon\|_{W^{-\theta,p}} \le \varepsilon,$$

and if we assume $f \in W^{s,p}$ with $\ell \ge s - 1$, the optimal estimate is in the order of $\varepsilon^{s/(s+\theta)}$, and achieved for $h \approx \varepsilon^{s/(s+\theta)}$. We first notice that if $\theta = 0$ (noise in L^2), this speed is of order ε : we do not need any regularisation, and we face a well-posed inverse problem!

We notice that this result is fully coherent with the statistical estimates of Propositions 4.1 and 4.2: to see it, the correct heuristics consists in taking $\theta = \frac{1}{2}$ (for a heuristics of the regularity of the empirical measure) and $\varepsilon = n^{-1/2}$ (the noise level being given by a central limit theorem), see Ref. [100] for an illuminating explanation of this comparison. We then have an estimate in the order of

$$\varepsilon^{s/(s+\theta)} = \varepsilon^{s/(s+1/2)} = n^{-s/(2s+1)},$$

as above. We further develop these heuristics or comparison between stochastic and deterministic noise in Sec. 4.2.2 below.

Finally, we note that the proof of Proposition 4.4 is not more involved for $k = 2$ than for $k = 1$, contrarily to the stochastic setting where the selection bias and the dependence between the individuals in the population case make it much more complex.

4.1.2. *Population point data*

Let us imagine that we are given a noisy measurement of the distribution of cells $N_k(a)$. This noise may be modeled by three different settings, increasingly realistic:

1. deterministic noise model: we model the noise by a measurement N_k^ε such that

$$\|N_k^\varepsilon - N_k\|_{W^{-s,p}([0,\infty))} \le \varepsilon.$$

2. Stochastic sampling noise: we assume that we observe a sample of ages $a_1, \cdots, a_n$ realisations of $A_1, \cdots, A_n$ i.i.d. random variables of density N_k. We could refine this setting by adding a measurement noise to each a_i.

3. Stochastic process: we observe, at a given time T, a sample of ages of cells. This means that we observe $\{a_u\}$, $u \in \mathcal{U}_3$ defined by

$$a_u = T - b_u, \qquad u \in \mathcal{U}_3 = \{u \in \mathcal{U}, \quad b_u \le T < b_u + \zeta_u\}.$$

For the third model, one first needs to establish asymptotic results as done in Ref. [14] given in Proposition 3.8 above for individual dynamics data. Having

$$\mathcal{E}(g) = 2\lambda_2 \int_0^\infty g(x) e^{-\lambda_2 x} \exp\left(-\int_0^x B(y) dy\right) dx$$

and ignoring the fact that λ_2 is unknown, we can anticipate that by picking a suitable test function g as a kernel, the information about $B(x)$ can only be inferred through $\exp(-\int_0^x B(y) dy)$. More precisely, consider the quantity

$$\widehat{f}_{h,T}(x) = -\mathcal{E}^T\left(\lambda_2 (K_h)'(x - \cdot)\right)$$

for a kernel satisfying (4.5). By Proposition 3.8 and integrating by part, we readily see that

$$\widehat{f}_{h,T} \mapsto -\mathcal{E}\left(\frac{1}{\lambda_2}(K_h)'(x - \cdot)\right) = \int_0^\infty K_h(x - y) f_{B+\lambda_2}(y) dy \qquad (4.15)$$

in probability as $T \to \infty$, where $f_{B+\lambda_2}$ is the density associated to the division rate $B(x) + \lambda_2$. On the one hand, using following line by line the proofs of Ref. [14] it is not difficult to show that the rate of convergence in (4.15) is of order $h^{-3/2} e^{\lambda_2 T/2}$ since we take the derivative of the kernel K_h. On the other hand, the limit $\int_0^\infty K_h(x - y) f_{B+\lambda_2}(y) dy$ approximates $f_{B+\lambda_2}(x)$ with an error of order h^s if B is s−Hölder. Balancing the two error terms in h, we see that we can estimate $f_{B+\lambda_2}(x)$ with an error of (presumably optimal) order $\exp(-\lambda_2 \frac{s}{2s+3}T)$. Due to the fact that the denominator in representation (4.10) can be estimated with parametric error rate $\exp(-\lambda_2 T/2)$ (possibly up to polynomially slow terms in T), we end up with the rate of estimation $\exp(-\lambda_2 \frac{s}{2s+3}T)$ for $B(x)$ as well, and that can be related to an ill-posed problem of order 1 (see for instance Ref. [96]).

This phenomenon, namely the structure of an ill-posed problem of order 1 in restriction to data alive at time T, appears in the other settings: for the estimation of a size-division rate from living cells at a given large time in Refs. [99, 101] or for the estimation of the dislocation measure for a homogeneous fragmentation, see Ref. [10].

For the first and second noise models, we use the explicit formula (3.5) to get

$$B(a) = -\lambda_k - \frac{\partial_a N_k(a)}{N_k(a)} \tag{4.16}$$

which is equivalent to (4.15). As for individual dynamics data, we see that we need to divide by the density, so that we will not be able to estimate B at places where it vanishes. The new fact is that, contrarily to the formula given by Eq. (4.13), the formula depends on the age-derivative of N_k, so that, as shown below in Proposition 4.5, the so-called *degree of ill-posedness* of the inverse problem is the one of estimating a function from its derivative: as for the third model, more regularisation is needed. We obtain the two following propositions.

Proposition 4.5: (Deterministic noise) *Under the assumptions of Proposition 4.4, defining*

$$H_{\varepsilon,h}(a) := -\lambda_k^{\varepsilon} N_k^{\varepsilon}(a) - (K_h * \partial_a N_k^{\varepsilon})(a), \qquad H(a) = B(a) N_k(a),$$

we have

$$\|H_{\varepsilon} - H\|_{L^p} \leq C(K, N_k) \left(\|K_h * N_k - N_k\|_{L^p} \right.$$

$$\left. + |\lambda_k^{\varepsilon} - \lambda_k| + h^{-\theta-1} \|N_k^{\varepsilon} - N_k\|_{W^{-\theta,p}} \right)$$

with $C(K, N_k)$ depending only on K, K' and N_k.

The proof is left to the reader; it is exactly the same ingredients as before, namely standard convolution inequalities. We see that for $N_k \in W^{s,p}$ and $\ell \geq s - 1$, and an error ε in $W^{-\theta,p}$, we have an optimal estimate in the order of $\varepsilon^{\frac{s}{s+\theta+1}}$, corresponding, for $\theta = 0$, to an inverse problem of degree of ill-posedness 1.

Proposition 4.6: (Stochastic sampling noise) *Under the assumptions of Proposition 4.5, assume that we know λ_k from previous observations, and that we observe an i.i.d. sample $a_1, \cdots, a_n$ of law N_k, and define the* *empirical measure*

$$N_n(da) = \frac{1}{n}\sum_{i=1}^{n}\delta_{a_i}(da)$$

and its regularisation

$$N_{n,h}(a)da = K_h * \left(\frac{1}{n}\sum_{i=1}^{n}\delta_{a_i}(da)\right) = \frac{1}{n}\sum_{i=1}^{n}K_h(a-a_i)da.$$

We define

$$H_{n,h}(a) := -\lambda_k - N_{n,h}(a) - \partial_a N_{n,h}(a), \qquad H(a) := B(a)N_k(a),$$

and the bias

$$\mathfrak{b}_h(N_k)(a) = \big|(K_h * N_k)(a) - N_k(a)\big|.$$

We have the following estimate

$$\mathbb{E}\left[\big|H_{n,h}(a) - H(a)\big|^2\right] \leq \mathfrak{b}_h(N_k)(a)^2 + C(N_k,K)\frac{1}{nh^3}, \qquad (4.17)$$

where $C(B,K)$ depends only on N_k, K and K'.

The proof is close to the proof of Proposition 4.1. The inflation in the variance term from the order $(nh)^{-1}$ to $(nh^3)^{-1}$ comes from the fact that we take a derivative $\partial_a N_{n,h}(a)$ of the kernel estimator $N_{n,h}(a)$. As previously seen, the bias is of order h^s if $N_k \in W^{s,p}$ and $\ell \geq s - 1$; this leads to an optimal error (*i.e.* the square root of the left-hand side in (4.17)) in the order of $n^{-\frac{s}{2s+3}}$. Taking in Proposition 4.5 $\theta = 1/2$ and $\varepsilon = n^{-1/2}$, we have $\varepsilon^{\frac{s}{s+3/2}} = n^{-\frac{s}{2s+3}}$: here again, this is the same optimal speed of convergence.

All these orders of magnitude for the convergence rates remain true for the size-structured model studied below.

4.2. *Estimating a size-dependent division rate*

We now review the methods and results developed to estimate a size-dependent division rate, as built in Sec. 2.3 and analysed in Sec. 3.3. We follow the same notations as above. We do not treat here the interesting question of estimating the fragmentation kernel $b(y,x)$, and refer for instance to Refs. [9, 10, 102, 103].

4.2.1. *Individual dynamics data*

Combining stochastic, deterministic and asymptotic approaches allows us to obtain easily reconstruction formulae.

Let us first revisit heuristically the formulae of Refs. [19, 37]. In general terms, we observe

$$\{(\xi_u, \chi_u, \zeta_u), \ u \in \mathcal{U}_k\}, \qquad k = 1 \text{ or } k = 2, \qquad \text{or} \qquad \{\xi_u^T, \ u \in \mathcal{V}_T\},$$

with (ξ_u, χ_u, ζ_u) respectively the size at birth, size at division and lifetime of the individual u taken in the sample $\mathcal{U}_k$ defined by (4.1) and (4.2), $\mathcal{V}_T$ defined by (3.8) and ξ_u^T the size of the individual $u \in \mathcal{V}_T$ alive at time T. $\mathcal{U}_1$ models the genealogical observation and individual dynamics data, $\mathcal{U}_2$ the population observation and individual dynamics data, and $\mathcal{V}_T$ population point data and population observation.

Assuming that the asymptotic behaviour of Sec. 3.3 has been reached in each of these models, we can interpret the densities with the help of Eq. (3.16).

- $\xi_u, u \in \mathcal{U}_k$, has a density distribution given by

$$f_k^b(x) = \frac{k \int_x^\infty B(y)\tau(y)b(y,x)N_k(y)dy}{k \int_0^\infty \int_s^\infty B(y)\tau(y)b(y,s)N_k(y)dyds} = \frac{\int_x^\infty B(y)\tau(y)b(y,x)N_k(y)dy}{\int_0^\infty B(y)\tau(y)N_k(y)dy}.$$

  This formula is obtained by identifying the last term in (3.16) with the newborn proportion, and normalise it to obtain a density.

- $\chi_u, u \in \mathcal{U}_k$, has density distribution given by

$$f_k^d(x) = \frac{B(x)\tau(x)N_k(x)}{\int_0^\infty B(y)\tau(y)N_k(y)dy}.$$

- ξ_u^T has for density N_k.

To estimate B, we can then write

$$B(x) = \frac{f_k^d(x)}{\tau(x)N_k(x)} \int_0^\infty B(y)\tau(y)N_k(y)dy,$$

and it remains to find formulae for τN_k and for $\int_0^\infty B(y)\tau(y)N_k(y)dy$.

Case $k = 1$ (genealogical observation). We denote $C = \int_0^\infty B(y)\tau(y)N_1(y)dy$ and write (3.16) as

$$\partial_x(\tau N_1) + Cf_1^d = Cf_1^b \implies \frac{\tau(x)N_1(x)}{C} = \int_x^\infty (f_1^d(y) - f_1^b(y)) \, dy,$$

so that we obtain the reconstruction formula

$$B(x) = \frac{f_1^d(x)}{\int_x^\infty \left(f_1^d(y) - f_1^b(y)\right) dy}. \tag{4.18}$$

Case $k = 2$ **(population observation).** We then have $\int_0^\infty B(y)\tau(y)N_2(y)dy = \lambda_2$, and (3.16) may be written as

$$\lambda_2 N_2 + \partial_x(\tau N_2) + \lambda_2 f_2^d = 2\lambda_2 f_2^b,$$

from which we deduce

$$\frac{\tau(x)N_2(x)}{\lambda_2} = \int_x^\infty \left(f_2^d(y) - 2f_2^b(y)\right) e^{\lambda_2 \int_x^y \frac{ds}{\tau(s)}} dy,$$

leading finally to the reconstruction formula

$$B(x) = \frac{f_2^d(x)}{\int_x^\infty \left(f_2^d(y) - 2f_2^b(y)\right) e^{\lambda_2 \int_x^y \frac{ds}{\tau(s)}} dy}. \tag{4.19}$$

Using either (4.18) or (4.19) and replacing f_k^d and f_k^b by the empirical distribution obtained from samples (ξ_i^b, ξ_i^d), we can estimate B without any knowledge on the fragmentation kernel b, as soon as both newborn and dividing cells distributions are observed. For genealogical data ($k = 1$), even the growth rate τ may be unknown. This remark may be generalised to other models, see for instance the formula (14) of Ref. [37] for a model with $k = 1$, the mitosis kernel (for which we have the simplification $f_k^b(x) = 2f_k^d(2x)$) and variable growth rates.

Individual dynamics data, stochastic approach, genealogical observation In the cases where we model the noise either deterministically or through an *i.i.d.* sample, the reconstruction formulae (4.18) and (4.19) immediately yield estimation results similar to the ones of Propositions 4.4 and 4.1 stated for the age model. More involved is the case where we do not depart from (3.16) but from the stochastic model; it has been studied in Ref. [37] for the case of genealogical observation, easier to study than the population observation case. To our best knowledge, the solution for population observation, with all the difficulties we already mentioned for the age problem (selection bias, censoring, non-ancillarity) plus a more intricate model, remains open.

In this setting, we look for a nonparametric estimator of $x \mapsto B(x)$ using the observation scheme $\mathcal{U}_1$ defined in Sec. 4.1.1. We thus observe

$$(\xi_u)_{u \in \mathcal{U}_1} \quad \text{and} \quad (\zeta_u)_{u \in \mathcal{U}_1}.$$

We have (see Sec. 2.1)

$$\mathbb{P}(\chi_u \in (x, x + dx) | \chi_u \ge x) = B(x)dx = \tau(x)B(x)dt,$$

$$\mathbb{P}(\chi_u \ge x | \xi_u) = \mathbb{1}_{\{x \ge \xi_u\}} \exp\left(-\int_{\xi_u}^x B(y)dy\right).$$

Hence we infer, taking the equal mitosis kernel so that $2\xi_u = \chi_{u^-}$,

$$\mathbb{P}(\xi_u \in (x', x' + dx') | \xi_{u^-} = x) = 2B(2x')\mathbb{1}_{\{2x' \ge x\}} \exp\left(-\int_x^{2x'} B(y)dy\right) dx'.$$

We thus obtain a simple and explicit representation for the transition kernel $\mathcal{P}_B(x, dx') = \mathcal{P}_B(x, x')dx'$ as

$$\mathcal{P}_B(x, x') = 2B(2x')\mathbb{1}_{\{2x' \ge x\}} \exp\left(-\int_x^{2x'} B(y)dy\right).$$

Under appropriate conditions on B set out in details below, there exists a unique invariant probability $\nu_B(dx) = \nu_B(x)dx$ on $[0, \infty)$ such that the following contraction property holds

$$\sup_{|g| \le V} \left| \mathcal{P}_B^k g(x) - \int_0^\infty g(z)\nu_B(z)dz \right| \le RV(x)\gamma^k \tag{4.20}$$

(where, for an integer $k \ge 1$, we set $\mathcal{P}_B^k = \mathcal{P}_B^{k-1} \circ \mathcal{P}_B$) for an appropriate Lyapunov function V and some (explicitly computable) $\gamma < 1$. The proof of (4.20) goes along a classical scheme and is detailed in Proposition 4 of Ref. [10], and for $\tau(x) = \kappa x$ (4.20) holds with

$$V(x) = \exp\left(\frac{m}{\kappa\mu}x^\mu\right)$$

for $\mu > 0$. Expand further the equation $\nu_B \mathcal{P}_B = \nu_B$:

$$\nu_B(y) = \int_0^\infty \nu_B(x)\mathcal{P}_B(x, y)dx$$

$$= 2B(2y) \int_0^{2y} \nu_B(x) \exp\left(-\int_x^{2y} B(y')dy'\right) dx$$

$$= 2B(2y) \int_0^\infty \int_0^\infty \mathbb{1}_{\{x \le 2y, y' \ge y\}} \nu_B(x) \mathcal{P}_B(x, y')dy'dx.$$

This yields the key representation

$$\nu_B(y) = 2B(2y)\mathbb{P}_{\nu_B}(\xi_{u^-} \le 2y, \, \xi_u \ge y).$$

We conclude

$$B(y) = \frac{1}{2} \frac{\nu_B(y/2)}{\mathbb{P}_{\nu_B}(\xi_u^- \le y, \xi_u \ge y/2)} \tag{4.21}$$

and this yields the estimator

$$\widehat{B}_{n,h}(y) = \frac{1}{2} \frac{n^{-1} \sum_{u \in \mathcal{U}_{[n]}} K_h(\xi_u - y/2)}{n^{-1} \sum_{u \in \mathcal{U}_{[n]}} \mathbf{1}\{\xi_{u-} \leq y, \xi_u \geq y/2\} \vee \varpi_n},$$

where the kernel $K_h(y) = h^{-1}K(h^{-1}y)$ is specified with an appropriate bandwidth (and technical threshold $\varpi_n > 0$).

We assess the quality of $\widehat{B}_n$ in squared-loss error over compact intervals $\mathcal{D}$. We need to specify local smoothness properties of B over $\mathcal{D}$, together with general properties that ensure that the empirical measurements converge with an appropriate speed of convergence. This amounts to impose an appropriate behaviour of B near the origin and infinity.

For $\alpha > 0$ and positive constants r, m, ℓ, L, introduce continuous functions $B : [0, \infty) \to [0, \infty)$ such that

$$\int_0^{r/2} x^{-1}B(2x)dx \leq L, \quad \int_{r/2}^r x^{-1}B(2x)dx \geq \ell, \quad B(x) \geq m\,x^\alpha \text{ for } x \geq r. \tag{4.22}$$

Define

$$\delta := \frac{1}{1 - 2^{-\alpha}} \exp\left(-(1 - 2^{-\alpha})\frac{m}{\kappa\alpha}r^\alpha\right).$$

Let γ denote the spectral radius of the operator $\mathcal{P}_B - 1 \otimes \nu_B$ acting on the Banach space of functions $g : [0, \infty) \to \mathbb{R}$ such that

$$\sup\{|g(x)|/V(x), x \geq 0\} < \infty.$$

Assumption 2: We have $\delta < \frac{1}{2}$ and $\gamma < 1$.

It is possible to obtain bounds on r, m, ℓ, L so that Assumption 2 holds, by using explicit bounds on γ following Hairer and Mattingly [104], see also Baxendale [105]. We are ready to state the performance of the estimator.

Theorem 4.7: *(Adapted from Theorem 2 from Ref. [37])* Specify $\widehat{B}_{n,h}$ with a kernel K satisfying Assumption 4.5 for some $n_0 > 0$ and

$$h = n^{-1/(2s+1)}, \quad \varpi_n \to 0.$$

For every compact interval $\mathcal{D} \subset (0, \infty)$ such that $\inf \mathcal{D} \geq r/2$, there exists a choice of m, ℓ, L, α such that Assumption 2 is satisfied. For B s-Hölder satisfying (4.22), we have

$$\mathbb{E}_\mu\left[\|\widehat{B}_n - B\|_{L^2(\mathcal{D})}^2\right]^{1/2} \lesssim \varpi_n^{-1} n^{-s/(2s+1)},$$

where $\mathbb{E}_\mu[\cdot]$ denotes expectation with respect to any initial distribution $\mu(dx)$ for $\xi_\emptyset$ on $(0, \infty)$ such that $\int_0^\infty V(x)^2\mu(dx) < \infty$.

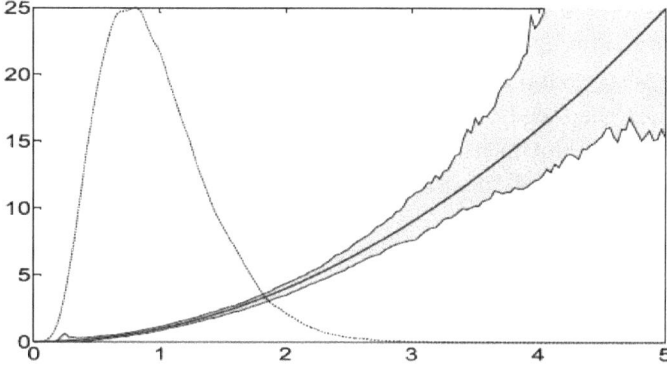

Fig. 9. Reconstruction for $n = 2^{10}$, error band for 95%, 100 simulations, with a small threshold $\varpi_n = 1/n$. We see that the error grows larger as x grows larger, as expected since the density of cells vanishes. Figure 3 from Ref. [37].

We illustrate this result in Fig. 9 for the reconstruction of a rate $xB(x) = x^2$. Since ϖ_n is arbitrary, we obtain the classical rate $n^{-s/(2s+1)}$ which is optimal in a minimax sense for density estimation. It is optimal in our context, using for instance the results of Ref. [106]. The knowledge of the smoothness s that is needed for the construction of $\widehat{B}_n$ is not realistic in practice. An adaptive estimator could be obtained by using a data-driven bandwidth in the estimation of the invariant density $\nu_B(y/2)$ in (4.21). In this context, the Goldenschluger-Lepski bandwidth selection method would presumably yield adaptation, see Ref. [99]. Finally, let us revisit the representation formula

$$B(y) = \frac{1}{2} \frac{\nu_B(y/2)}{\mathbb{P}_{\nu_B}\left(\xi_{u^-} \leq y, \xi_u \geq y/2\right)}$$

by noticing that we always have $\{\xi_{u^-} \geq y\} \subset \{\xi_u \geq y/2\}$, hence

$$\mathbb{P}_{\nu_B}\left(\xi_{u^-} \leq y, \xi_u \geq y/2\right) = \mathbb{P}_{\nu_B}\left(\xi_u \geq y/2\right) - \mathbb{P}_{\nu_B}\left(\xi_{u^-} \geq y\right)$$

$$= \int_{y/2}^{\infty} \nu_B(x)dx - \int_{y}^{\infty} \nu_B(x)dx$$

$$= \int_{y/2}^{y} \nu_B(x)dx.$$

Finally, for constant growth rate, we obtain

$$B(y) = \frac{1}{2} \frac{\nu_B(y/2)}{\int_{y/2}^{y} \nu_B(x)dx},$$

where we recognize (4.18). The "gain" in the convergence rate $n^{-s/(2s+1)}$ versus the rate $n^{-s/(2s+3)}$ obtained in the proxy model based on the transport-fragmentation equation, as seen in Sec. 4.1 for the age model, comes from the fact that we estimate the invariant measure "at division" versus the invariant measure "at fixed time" in the proxy model. In other words, there is more "nonparametric statistical information" in data extracted from $\mathcal{U}_2$ rather than $\mathcal{V}_T$, regardless of the tree structure. However $|\mathcal{V}_T| \approx |\mathcal{U}_2|$ as stems from the supercritical branching process structure.

4.2.2. *Population point data*

We now turn to the case of less rich observation schemes, where we only have access to population point data, which relate to the eigenvector solution N_k. For the noise, we consider either a deterministic noise or a sampling noise: as for the individual dynamics case, a complete approach, which would depart from the stochastic branching tree, has not yet been carried out.

Contrarily to the case of individual dynamics data, the shape of the fragmentation kernel is important here, and needs to be previously known, as well as the Malthusian parameter λ_k and the growth rate τ.

Estimating the division rate B may be formulated by the following inverse problem.

We know *a priori* the growth rate τ, the fragmentation kernel $b(y,x) = \frac{1}{y}b_0(\frac{x}{y})$ and the Malthusian parameter λ_k, and measuring either N_k^ε (deterministic noise) such that

$$\|N_k^\varepsilon - N_k\|_{W^{-s,p}([0,\infty))} \leq \varepsilon,$$

or a sample $x_1, \cdots, x_n$ realisations of $X_1, \cdots, X_n$ *i.i.d.* random variables of law N_k (sampling noise), with (N_k, λ_k) solution to (3.16) *i.e.*

$$\begin{cases} \frac{\partial}{\partial x}(\tau N_k(x)) + \lambda_k N(x) = -(\tau B N_k)(x) + k\int_0^1 (\tau B N_k)(\frac{x}{z})\frac{b_0(dz)}{z}, \\ \tau N_k(0) = 0. \end{cases}$$

$$(4.23)$$

How to estimate B?

We notice the same ingredients as for the age model: $1/$ B appears in the equation only multiplied by τN_k, so that we cannot estimate it at places where τN_k vanishes; $2/$ the equation reveals a size-derivative of N_k, so that we expect an inverse problem of degree of ill-posedness one as in Sec. 4.1.2; 3/ replacing the unknown B by the density of dividing cells BN_k, we transform a nonlinear inverse problem into a linear one.

However, contrarily to the simple age model, we do not have any explicit formula here: a dilation operator needs to be inverted. We decompose (4.23) into

$$L(N_k) = G_k(B\tau N_k) \quad \text{with} \quad G_k(f)(x) := k \int_0^1 f(\tfrac{x}{z}) \tfrac{b_0(dz)}{z} - f(x),$$

$$(4.24)$$

$$L(N_k) := \partial_x(\tau N_k) + \lambda_k N_k,$$

and estimate B through the following steps:

- Solve $G_k(f) = L$ for f, L in suitable spaces: for simplicity of inversion formulae, we thus choose Hilbert spaces of type $L^2([0, \infty), x^p dx)$ [107]. Note that under this shape, the problem to solve $N_k \to f = B\tau N_k$ is now linear.
- Estimate $L(N_k)$ from the measurement in the chosen space: to do so, we can either depart from N_k^ε or from a statistical sample, and then divide by (an estimate of) N_k and threshold appropriately.

The second step is exactly the same as what has been done in Sec. 4.1 for the renewal equation, so that we will not detail this part anymore here, and focus on the first step, which is specific to the growth-fragmentation equation.

Solving a dilation equation

The first step consists in inverting the dilation operator G_k defined by (4.24). We begin to treat the equal mitosis case, first studied by B. Perthame and J. Zubelli [12, 99, 101], and then turn to more general fragmentation kernels [108].

For the equal mitosis or diagonal kernel $b_0(x) = \delta_{x=\frac{1}{2}}$, we have the following result.

Proposition 4.8: (Adapted from Theorem A.3 of Ref. [101]) *Let G_k defined by (4.24), $b_0(x) = \delta_{x=\frac{1}{2}}$, $L \in L^2(x^p dx)$ with $p \neq 2k - 1$.*

There exists a unique solution $f \in L^2(x^p dx)$ to

$$G_k(f) = 2kf(2\cdot) - f(\cdot) = L,$$

and this solution depends continuously on $\|L\|_{L^2(x^p dx)}$. Moreover, defining

$$H_k^0 := \sum_{j=1}^\infty (2k)^{-j} L(2^{-j} x), \qquad H_k^\infty := -\sum_{j=0}^\infty (2k)^j L(2^j x),$$

we have $f = H_k^0$ if $p < 2k - 1$ and $f = H_k^\infty$ if $p > 2k - 1$. If $L \in L^q$ then $H_0 \in L^q$ for any $1 \leq q \leq \infty$.

For $L = 0$, any distribution of the form $f(\frac{\ln x}{x^2})$ with $f \in \mathcal{D}'([0, \infty))$ $\ln -2$ periodic is solution.

Proof: The proof is based on the Lax-Milgram theorem applied to the bilinear forms

$$\begin{cases} a_p(u, v) = \int \left(-2ku(2x) + u(x)\right) v(x) x^p dx, \\ \\ b_p(u, v) = \int \left(2ku(2x) - u(x)\right) v(2x) x^p dx, \end{cases}$$

and by Cauchy-Schwarz inequality, a_p and b_p are respectively coercive for $p > p_k$ and $p < p_k$ with $p_1 = 1$ and $p_2 = 3$.   □

In this result, we first notice that uniqueness depends on the space chosen: there is no reason, generally speaking, to have $H_k^0 = H_k^\infty$, so that departing from an estimate $L(N_k^\varepsilon)$, we have infinitely many choices. We first restrict to the two solutions H_k^0 and H_k^∞, since the others do not vanish fast enough at infinity compared to the one we look for, recall Theorem 3.9. Among these two solutions, H_k^0 "behaves better" at infinity, and H_k^∞ at zero, hence in Ref. [108] we proposed a combination of both solutions as a best approximation in $L^2((x^p + 1)dx)$ for $p > 3$, namely, for a given L, we define

$$H_k^{\bar{x}} = H_k^0 \mathbb{1}_{\{x \leq \bar{x}\}} + H_k^\infty \mathbb{1}_{\{x > \bar{x}\}}. \tag{4.25}$$

This solution is no more an exact solution of $G_k(f) = L$ unless $H_k^0 = H_k^\infty$. However, this property being satisfied for the "true" underlying distribution, it defines a convenient approximation as shown below.

For general self-similar fragmentation kernels, we use the Mellin transform — an important tool for the study of the equation in many cases, see Refs. [64, 83]. We recall that the Mellin transform $\mathcal{M}$ is an isometry between $L^2(x^q dx)$ and $L^2(\frac{q+1}{2} + i\mathbb{R})$ defined by

$$\mathcal{M}[f](s) := \int_0^\infty x^{s-1} f(x) dx, \mathcal{M}_q^{-1}[F](x) := \int_{-\infty}^\infty x^{-\frac{q+1}{2} - iv} F\left(\frac{q+1}{2} + iv\right) dv. \tag{4.26}$$

The Mellin transform for the operator G_k is then

$$\mathcal{M}[\mathcal{G}(f)](s) = (k\mathcal{M}[b_0](s) - 1)\mathcal{M}[f](s),$$

and we see that, if $|k\mathcal{M}[b_0](s) - 1|$ is bounded from below by a positive constant on the integration line $\frac{q+1}{2} + i\mathbb{R}$, we can define the inverse

$$H_k^q := \mathcal{M}_q^{-1} \left[\frac{\mathcal{M}[\mathcal{G}(f)](s)}{k\mathcal{M}[b_0](s) - 1} \right]. \tag{4.27}$$

We notice that for $s_k = k$ we have $k\mathcal{M}[b_0](s_k) - 1 = 0$. This zero, together with the isometry given by (4.26) and the inversion formula (4.27), gives insight on why $p_k = 2k - 1$ is the pivot in Proposition 4.8 (we have $s_k = \frac{p_k+1}{2}$), and why $H_k^q \neq H_k^{q'}$ if $q < s_k < q'$: the residue theorem quantifies exactly their difference, see Proposition 4 in Ref. [108] for more details.

Estimate with a deterministic noise Let us assume here that we measure N_k up to a deterministic noise of level $\varepsilon > 0$, $0 \leq \theta < 1$:

$$\|N_k - N_k^\varepsilon\|_{H^{-\theta}([0,\infty))} \leq \varepsilon.$$

By the general theory of linear inverse problems, let us pick any regularisation method of optimal order [107], of parameter $h > 0$, and define an approximation $L(N_k^\varepsilon)_h$ such that, for $N_k \in H^m([0, \infty))$, and $q > 2k - 1$, we have

$$\|L(N_k^\varepsilon)_h - L(N_k)\|_{L^2((1+x^q)dx)} \leq C \left(\frac{\varepsilon}{h} + h^m \right),$$

with C depending only on the method chosen and on the norm of N_k in H^m. Since we want to estimate $H_k = BN_k$ in $L^2((1 + x^q)dx)$ with $q > 2k - 1$, we define for some $a > 0$

$$H_{\varepsilon,h} := \mathcal{M}_0^{-1} \left[\frac{\mathcal{M}[L(N_k^\varepsilon)_h](s)}{k\mathcal{M}[b_0](s) - 1} \right] \mathbb{1}_{\{x \leq a\}} + \mathcal{M}_q^{-1} \left[\frac{\mathcal{M}[L(N_\varepsilon)_h](s)}{k\mathcal{M}[b_0](s) - 1} \right] \mathbb{1}_{\{x > a\}} \tag{4.28}$$

and we get the following proposition.

Proposition 4.9: (Adapted from Ref. [108], Theorem 1.1) *For* $N \in H^m([0, \infty))$ *solution to the eigenequation* (4.23) *we have*

$$\|N - N_\varepsilon\|_{H^{-\theta}([0,\infty))} \leq \varepsilon \implies \|H_{\varepsilon,h} - BN_k\|_{L^2((1+x^q)dx)} \leq C \left(\frac{\varepsilon}{h^{1+\theta}} + h^m \right)$$

where C *depends only on* $\|N\|_{H^m}$ *and the regularisation method chosen.*

Estimation with a stochastic noise Let us now assume that we observe a sample of n cells, of sizes $x_1, \cdots, x_n$, that are realisations of

$$(X_1, \ldots, X_n),$$

where the X_i are independent, with common density distribution $N(x)dx$ (recall that we have $N \geq 0$ and that we pick the normalisation $\int_0^\infty N = 1$). From (4.24), we have the formal representation

$$B = \frac{(G_k)^{-1}(L(N_k))}{\tau N_k}.$$

Thus, from data $(X_1, \ldots, X_n)$ we can build a regularisation

$$\widehat{N}_{h,k} = \frac{1}{nh} \sum_{i=1}^n K_h(\cdot - X_i)$$

which simply amounts to have a convolution of the empirical measure $n^{-1} \sum_{i=1}^n \delta_{X_i}(dx)$ with a well-behaved kernel $K_h(\cdot) = h^{-1}K(h^{-1}\cdot)$. The resulting $\widehat{N}_{h,k}$ being smooth (by picking K smooth enough), we may compute the action of the differential operator L on $\widehat{N}_h$. As soon as the operator G_k has bounded inverse, or that a nice approximation $(\mathfrak{G}_k)^{-1}$ of $(G_k)^{-1}$ is available, we may form the simple estimator:

$$\widehat{B}_{n,h} = \frac{(\mathfrak{G}_k)^{-1}(L(\widehat{N}_{h,k}))}{\tau \widehat{N}_{h,k}}. \tag{4.29}$$

In Ref. [99], we realise this program for the binary fragmentation operator and we propose a method to automatically select the bandwidth h from data. The following kind of results can be obtained: for a compact set $\mathcal{D}$, one can construct an approximation $(\mathfrak{G}_k)^{-1}$ and select a bandwidth h such that if $B \in H^s$, then

$$\mathbb{E}\left[\|\widehat{B}_{n,h} - B\|_{L^2(\mathcal{D})}\right] \lesssim n^{-s/(2s+3)}. \tag{4.30}$$

We obtain a rate of convergence that corresponds to an ill-posed problem of order 1, and this is consistent with the other approaches based on large population data (here of size n) alive at a given fixed, (but large) time.

Comparing deterministic and stochastic methods

The stochastic method rate of convergence $n^{-s/(2s+3)}$ for ill-posed problems of degree 1 in the statistical minimax theory [97] is to be compared with the deterministic method rate $\epsilon^{s/(s+1)}$, as stems from the classical theory exposed for instance in the classical textbook [107]. These are actually the same results, as stems from a classical analysis of comparison between stochastic and deterministic ill-posed inverse problems, see the illuminating paper [100].

Suppose we have an approximate knowledge of N and λ up to deterministic errors $\zeta_1 \in L^2$ and $\zeta_2 \in \mathbb{R}$ with noise level $\varepsilon > 0$: we observe

$$N_\varepsilon = N + \varepsilon\zeta, \quad \|\zeta\|_{L^2} \leq 1, \tag{4.31}$$

and

$$\lambda_\varepsilon = \lambda + \varepsilon\zeta_2, \quad |\zeta_2| \leq 1. \tag{4.32}$$

From the formal representation

$$B = \frac{(G_k)^{-1}\big(L(N_k)\big)}{\tau N_k},$$

the recovery of $L(N_k)$ is ill-posed in the terminology of Wahba [109] since it involves the computation of the derivative of N. If G_k is bounded with an inverse bounded in L^2 and the dependence in λ is continuous, the overall inversion problem is ill-posed of degree $a = 1$. By classical inverse problem theory for linear cases (here the problem is non-linear), this means that if $N \in W^{s,2}$, the optimal recovery rate in L^2-error norm should be $\varepsilon^{s/(s+a)} = \varepsilon^{s/(s+1)}$ (see also Refs. [12, 101]).

Suppose now that we replace the deterministic noise ζ_1 by a random Gaussian *white noise*: we observe

$$N_\varepsilon = N + \varepsilon\dot{W} = A\tfrac{\partial}{\partial x}N(x)dx + \varepsilon\dot{W} \tag{4.33}$$

where $\dot{W}$ is a Gaussian white noise (a random distribution) and $A\varphi(x) = \int_0^x \varphi(y)dy$ denotes the integration operator (which has degree of ill-posedness $a = 1$). This setting is actually statistically (asymptotically) very close to observing $(X_1, \ldots, X_n)$ where the X_i are independent random variables, with common density N, at least over compact intervals $\mathcal{D}$ and as soon as N does not degenerate, as follows from the celebrated result of Nußbaum [110].

In this setting, we want to recover $\tfrac{\partial}{\partial x}N$. Integrating, we equivalently observe

$$Y_\varepsilon(\cdot) = \int_0^\cdot A\tfrac{\partial}{\partial x}N(x)dx + \varepsilon\mathcal{W},$$

where $(\mathcal{W}_x, x \geq 0)$ is a Brownian motion. Applying formally the $1/2$-fractional derivative operator $D^{1/2}$, we recast the observation Y_ε into an equivalent observation

$$Z_\varepsilon = D^{1/2}\int_0^\cdot A\tfrac{\partial}{\partial x}N(x)dx + \varepsilon D^{1/2}\mathcal{W},$$

so that $D^{1/2}\mathcal{W}\cdot$ is in L^2. The operator $\varphi \mapsto D^{1/2}\int_0^\cdot A\varphi(x)dx$ maps L^2 onto $W^{1+1-1/2,2}$ *i.e.* we have an effective degree of ill-posedness $a = 3/2$. We should then obtain an optimal rate of the form

$$\varepsilon^{s/(s+3/2)} = \varepsilon^{2s/(2s+3)} = n^{-s/(2s+3)}$$

for the calibration $\varepsilon = n^{-1/2}$ dictated by (4.32) when we compare our statistical model with the deterministic perturbation. This is exactly the rate we find in (4.30): the deterministic error model and the statistical error model coincide to that extent.

4.3. Estimating an increment-dependent division rate

Individual dynamics data: a simple renewal problem?

Despite the intricate character of the adder model, which combines the influences of the age and size, in the case where one is given individual dynamics data, the estimate to do is exactly the same as for the pure age-structured model... at least for the genealogical observation case. In this case, given that we observe $a_1, \cdots, a_n$ increments at division, they form a renewal process, observed without bias, and we can apply (4.4) and standard density estimation methods as explained in Sec. 4.1.

For the population observation case ($k = 2$), things are more intricate: as for the previous models, there exists a selection bias, which is not the same as given by (4.13) since the growth rate of the increment is no more constant and moreover depends on size. However, in the case of exponential growth $\tau(x) = \kappa x$, we have the small miracle already mentioned that $\lambda_2 = \kappa$ and, with $C, C_d > 0$ normalisation constants, we have

$$N_1(a, x) = CxN_2(a, x) \implies f_1(a, x) = C_d x f_2(a, x),$$

where we denote $f_k(a, x)$ the density of dividing cells of increment a and size x. We can thus take advantage of Eq. (4.4) and write, for $f_k^a(a) = \int_0^\infty f_k(a, x)dx$ the marginal density along a:

$$B(a) = \frac{f_1^a(a)}{\int_a^\infty f_1^a(s)ds} = \frac{\int_0^\infty f_2(a, x)xdx}{\int_a^\infty \int_0^\infty f_2(s, x)xdxds}. \tag{4.34}$$

Instead of estimating directly the density $f_2^a(a)$ from the sample $(a_1, \cdots, a_n)$, as done for the genealogical case, one has to estimate a weighted density: from a sample $((A_1, X_1); \cdots (A_n, X_n))$ of increment and size at division of cells according a population observation, we define the following estimate for the debiased density f_1:

$$\hat{f}_1^a(a)da = K_h * \left(\frac{1}{n} \sum_{i=1}^{n} X_i \delta_{A_i}(da) \right) = \frac{1}{n} \sum_{i=1}^{n} X_i K_h(a - A_i)da, \quad (4.35)$$

for which we can prove the same rate of convergence as done for instance in Proposition 4.4.

Population point data: a severely ill-posed problem

When we have treated the case of population point data for the age-structured equation, in Sec. 4.1, we have remarked that it was rather an instructive toy model than a really interesting case, since if we are able to observe ages in a population, this should mean that we are also able to observe newborn among them, hence dividing cells, hence we would be back to the individual dynamics data.

In the case of the size-structured model, studied in Sec. 4.2.2 on the contrary, the inverse problem setting appears fully relevant in many experimental cases and different applications — fragmenting polymers, *in vivo* dividing cells, etc.

For the increment model, as for the renewal model, a "naive" inverse problem, which is to assume that we measure samples distributed along a density $N(a, x)$ solution to (3.19) seems irrelevant: observing samples distributed along $N(a, x)$ implies that we also observe newborn, *i.e.* $N(0, x)$, and then why not as well dividing cells — which would bring us back to the individual dynamics data, discussed in the paragraph above.

Much more realistic is the following case, studied in Ref. [111]: Assume we are given $x_1, \cdots, x_n$ a $n-$ sample of realizations of $X_1, \cdots, X_n$ *i.i.d.* random variables of density $N_k^x(x)$ defined by

$$N_k^x(x) := \int_0^\infty N_k(z, x)dz,$$

other said, N_k^x is the x-marginal, or yet the size-distribution of a sample of cells following an increment-structured dynamics, and having reached their steady behaviour given by $N_k(z, x)$. How to estimate an *increment*-dependent division rate $B(a)$ from a *size*-dependent distribution?

Surprisingly, the answer is given by the following proposition, which provides us with a reconstruction formula — and shows that this inverse problem is severely ill-posed.

Proposition 4.10: *(Adapted from Proposition 1 in Ref. [111]) We have the following reconstruction formula, where $\mathcal{F}$ and $\mathcal{F}^{-1}$ denote the*

Fourier and inverse Fourier transform:

$$B(z) = \frac{f(z)}{\int_z^\infty f(s)ds}, \qquad f(z) := \mathcal{F}^{-1}\left(\frac{\mathcal{F}[H_1(\cdot)]}{\mathcal{F}[H_1(2\cdot)]}\right),$$

where $H_1(x) = \kappa x \int_0^\infty B(z)x^{k-1}N_k(z,x)dz$ *is the solution of the dilation equation given in Proposition 4.8:*

$$\mathcal{L}(x) = \frac{\partial}{\partial x}(\kappa x^k N_k) = 2H_1(2x) - H_1(x). \tag{4.36}$$

From this formula, we deduce that if we observe $X_1, \cdots, X_n$ *an i.i.d. sample of law* $N_k^x(x)$, *we propose the following estimator for* B:

$$\widehat{B}_{n,h,h'}(z) = \frac{\widehat{f}_{n,h}(z)}{\widehat{S}_{n,h}(z)} = \frac{\int_{-1/h'}^{1/h'} \frac{\widehat{H_{1,n,h}(\cdot)^*}(\xi)}{\widehat{H_{1,n,h}(\cdot)^*}(\xi/2)} e^{-\mathrm{i}a\xi} d\xi}{\int_z^\infty \int_{-1/h'}^{1/h'} \frac{\widehat{H_{1,n,h}(\cdot)^*}(\xi)}{\widehat{H_{1,n,h}(\cdot)^*}(\xi/2)} e^{-\mathrm{i}s\xi} d\xi ds}$$

where $\widehat{H}_{k,n}$ *denotes an approximate solution to the dilation equation (4.36) as seen in Proposition 4.9 with, for the left-hand side,*

$$\mathcal{L}_{n,h}(x) := \frac{1}{n}\sum_{i=1}^n \frac{\partial}{\partial x}\left(\kappa x^k K_h(x - X_i)\right).$$

Proof: As a sketch of the proof in this simpler case where $\tau(x) = \kappa x$ (see Ref. [111] for a general growth rate $\tau(x)$) we first write the equation for N_1, noticing that $N_1 = x^{k-1}N_k$ up to a constant, and integrate it along z to find the dilation equation (4.36). We then solve the equation for $C(z,x) = \kappa x N_1(z,x)$ along the characteristics, and find

$$\kappa x N_1(z,x) = \kappa(x-z)N_1(0, x-z)e^{-\int_0^z B(s)ds}.$$

We use this expression in the definition of H_1 and find

$$H_1(x) = \int_0^x B(z)\kappa x N_1(z,x)dz = \int_0^x B(z)\kappa(x-z)N_1(0, x-z)e^{-\int_0^s B(s)ds}dz$$

using the boundary condition, we also have

$$\kappa x N_1(0,x) = 4\kappa x \int_0^{2x} B(z)N_1(z,2x)dz = 2H_1(2x)$$

hence

$$H_1(x) = \int_0^x B(z)e^{-\int_0^z B(s)ds}2H_1(2(x-z))dz = f_1 * (2H_1(2\cdot))(x)$$

which appears as a deconvolution problem, where $2H_1(2x)$ plays the role of "noise". $\qquad\square$

We refer to Ref. [111] for a thorough numerical investigation.

5. Application to experimental data

In the previous section, we have developed methods to estimate the division rate in three different models: age-structured, size-structured or size-increment-structured model. In the case where data are rich enough, these methods can be used not only to estimate the division rate but also to select which model is more likely. This model selection is carried out here on the case of individual dynamics data: in the case of point population data, estimating the division rate is possible, but the comparison between models is not, since the data is not informative enough.

5.1. *Guideline of a protocol*

To test a given (age, size or size-increment — or anything else not included in this chapter) model, we

- calibrate it as done in Sec. 4,
- simulate an age-size or increment-size model, in which our modelling assumption is embedded: see in Sec. 3.4 the model Eqs. (2.12) and (2.13). For instance, if we want to compare the adder to the sizer without generalising any assumption on the growth rate or on the fragmentation kernel, we can simulate the following model:

$$\frac{\partial}{\partial t} n_k + \frac{\partial}{\partial z}(\kappa x n_k) + \frac{\partial}{\partial x}(\kappa x n_k) = -\kappa x B(z, x) n_k(t, z, x),$$

$$n_k(t, z = 0, x) = 4k \int_0^\infty B(z, 2x) n_k(t, z, 2x) dz$$

till its asymptotic steady behaviour $n_k(t, a, x) = e^{\lambda t} N_k(a, x)$ is reached. This simulation step may be carried out either by using the PDE and an adequate numerical scheme, see for instance Refs. [89, 112], or by Monte-Carlo simulations.
- Compare quantitatively data and simulations, by defining a convenient distance between measures. This choice has always some intrinsic arbitrariness and is open to debate. We refer to the paper [113] that discusses Wasserstein distances (and some of their weighted variants), interpreted as connections from univariate methods like the Kolmogorov-Smirnov test, QQ plots and ROC curves, to other multivariate tests. A thorough discussion of statistical tests or distance choices in this context lies beyond the level of generality intended here. We nevertheless refer to Ref. [19] where an operational protocol is proposed to compare different division rate models.

5.2. *Some results*

This protocol has been used in Ref. [19] to compare the age-structured and the size-structured equation, compared quantitatively by L^2 norms between regularised solutions. We reproduce in Fig. 10 the sensitivity analysis carried out, which concluded that taking into account the experimental variability in the growth rates (recall Fig. 7(Left)) or in the fragmentation kernel b_0 (recall Fig. 7(Right)) does not improve significantly the results.

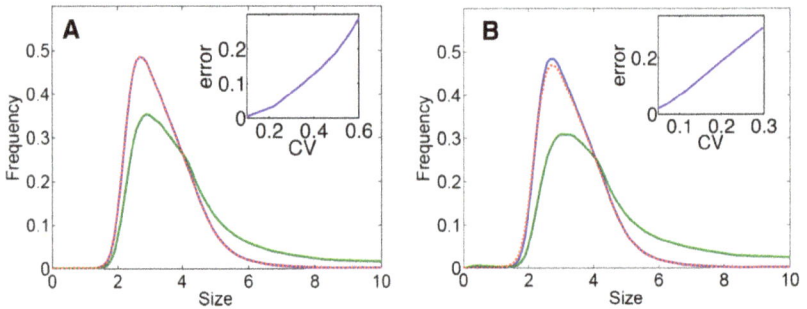

Fig. 10.   Effect of adding variability on the size-distribution of cells. A: adding variability in the growth rate, B: in the fragmentation kernel. In green the size distribution with a coefficient of variation of 60%, in dotted pink the simulation without variability, in blue the estimate of the experimental distribution. The insets show the distance between the distribution with no variability and the distance with a variability of given CV: the conclusion is that, for these experiments on bacteria, variability is negligible and the simplified models fit well. Figure 4 from Ref. [19].

This protocol has then been improved to take into account the adder model, and implemented in the prototype CellDivision plateform `https://celldivision.paris.inria.fr/welcome/`, developed by Adeline Fermanian [114]. This plateform estimates the three models — age, size or increment structured — when the user provides individual dynamics data of lifetimes, size at birth, size at division and/or increment of size at division, together with a growth law. We show on Fig. 11 an application on data from Ref. [5] also used in Ref. [19], for which it clearly appears that the incremental model fits better.

Of note, a very efficient way of comparing model and data, used in many biophysical papers such as Refs. [21, 115], consists in comparing the correlation coefficients between size, age and increment for the various models, since an age model predicts no correlation between size at birth and generation time, and an adder model between size at birth and increment at

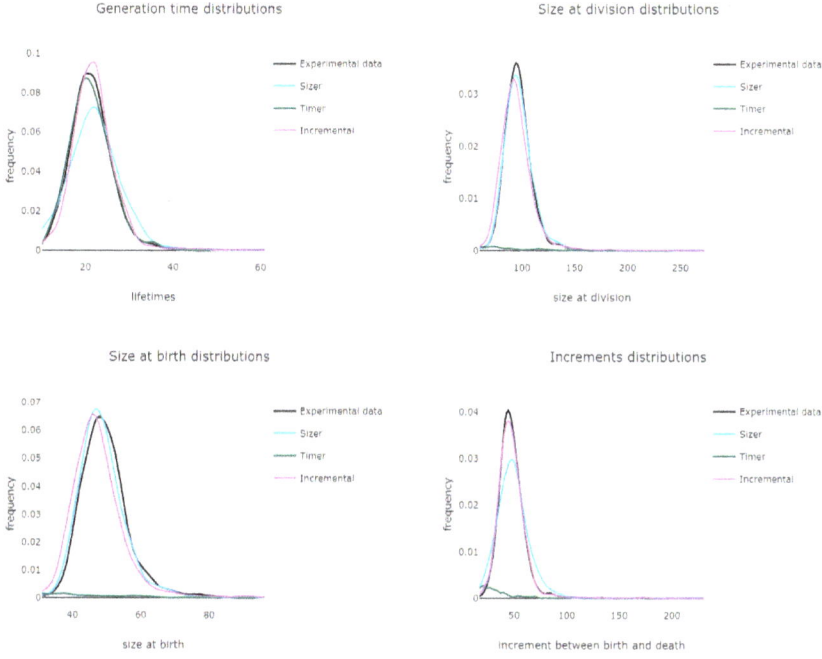

Fig. 11. Screenshot of results obtained by the prototype plateform CellDivision, developed by Adeline Fermanian, to fit genealogical data of bacterial division from Ref. [5].

division. We reproduce in Table 1 the results obtained on genealogical data from Ref. [5], comparing experimental correlations with the ones obtained from the calibrated incremental model (using the CellDivision plateform): the match is excellent.

Table 1. Correlation coefficients (C.C.) for the data shown in Fig. 11.
AD: Age at Division, SB: Size at Birth, SD: size at Division, ID: Increment at Division. Computed by the CellDivision prototype plateform [114].

C.C. between:	AD/SB	AD/SD	AD/ID	SB/SD	SB/ID	SD/ID
Experimental	-0.47	0.53	0.82	0.42	-0.03	0.89
Timer model	-0.02	0.04	0.08	0.98	0.93	0.98
Sizer model	-0.66	0.67	0.92	0.08	-0.39	0.89
Adder model	-0.48	0.51	0.86	0.49	-0.01	0.87

6. Perspectives and open questions

In this chapter, we retraced the methods developed by our group during the last decade, originating in Ref. [12], to estimate the division rate in linear structured population equation models. They crucially rely on the mathematical analysis of the long-term behaviour of these equations and processes, sketched in Sec. 3. Many developments still need to be done.

These models are well-adapted to model experiments in a steady environment, typically, with unrestricted nutrient and space. How to estimate the division in a non-steady environment? This could be of tremendous importance to understand for instance how cells react to an external stress, or to take into account cell-to-cell interaction.

We have focused in the chapter on three main models: age-structured, size-structured, and increment-structured model. The method is always the same, but the mathematical analysis of both the time asymptotics and of the inverse problem is specific. It would be very interesting to adapt it to richer models, for instance taking into account heterogeneity, or the G1/S/G2/M steps in the eucaryotic cell cycle, or yet models structured in DNA content, as the initiation of DNA replication appears to be the true important stage in the life of yeast cells [115] for instance.

The inference methods outlined in Sec. 4 could be enriched by taking into account other types of noise, in particular adding experimental measurement noise linked to image analysis. They could also lead to the design and study of statistical tests — especially interesting when the data is richer, given by individual dynamics data. In this case, we could also investigate the question of model selection: instead of depending on an *observed* variable, the division rate could depend on a *hidden* variable, more or less in the spirit of Sec. 4.3 for population point data [111].

Acknowledgments

The content of this chapter is based on a lecture given at IMS of National University of Singapore in January 2020. The authors thank Professor Weizhu Bao and IMS for providing such a nice opportunity. This work was partially supported by the ERC Starting Grant SKIPPERAD (Grant number 306321).

References

1. M.-C. Duvernoy, T. Mora, M. Ardré, V. Croquette, D. Bensimon, C. Quilliet, J.-M. Ghigo, M. Balland, C. Beloin and S. Lecuyer, Asymmetric adhesion of rod-shaped bacteria controls microcolony morphogenesis, *Nature communications.* **9** (2018), 1–10.
2. D. Dell'Arciprete, M. Blow, A. Brown, F. Farrell, J. S. Lintuvuori, A. McVey, D. Marenduzzo, and W. C. Poon, A growing bacterial colony in two dimensions as an active nematic, *Nature communications.* **9** (2018), 1–9.
3. M. Doumic, S. Hecht, and D. Peurichard, A purely mechanical model with asymmetric features for early morphogenesis of rod-shaped bacteria microcolony, *arXiv preprint arXiv:2008.04532.* (2020).
4. N. Q. Balaban, J. Merrin, R. Chait, L. Kowalik, and S. Leibler, Bacterial persistence as a phenotypic switch, *Science.* **305** (2004), 1622–1625.
5. P. Wang, L. Robert, J. Pelletier, W. L. Dang, F. Taddei, A. Wright, and S. Jun, Robust growth of Escherichia coli, *Curr. Biol.* **20** (2010), 1099–103.
6. A. Meunier, F. Cornet, and M. Campos, Bacterial cell proliferation: from molecules to cells, *FEMS Microbiology Reviews.* **45** (2021), fuaa046.
7. W.-F. Xue and S. E. Radford, An imaging and systems modeling approach to fibril breakage enables prediction of amyloid behavior, *Biophys. Journal.* **105** (2013), 2811–2819.
8. D. M. Beal, M. Tournus, R. Marchante, T. J. Purton, D. P. Smith, M. F. Tuite, M. Doumic, and W.-F. Xue, The division of amyloid fibrils: Systematic comparison of fibril fragmentation stability by linking theory with experiments, *Iscience.* **23** (2020), 101512.
9. M. Tournus, M. Escobedo, W.-F. Xue, and M. Doumic, Insights into the dynamic trajectories of protein filament division revealed by numerical investigation into the mathematical model of pure fragmentation. (2021), hal-03131754.
10. M. Hoffmann and N. Krell, Statistical analysis of self-similar conservative fragmentation chains, *Bernoulli.* **17** (2011), 395–423.
11. B. Basse, B. Baguley, E. Marshall, W. Joseph, B. van Brunt, G. Wake, and D. J. N. Wall, A mathematical model for analysis of the cell cycle in cell lines derived from human tumors, *J. Math. Biol.* **47** (2003), 295–312.
12. B. Perthame and J. Zubelli, On the inverse problem for a size-structured population model, *Inverse Problems.* **23** (2007), 1037–1052.
13. M. Escobedo, S. Mischler and M. Rodriguez Ricard, On self-similarity and stationary problem for fragmentation and coagulation models, *Annales de l'Institut Henri Poincare (C) Non Linear Analysis.* **22** (2005), 99–125.
14. M. Hoffmann and A. Olivier, Nonparametric estimation of the division rate of an age dependent branching process, *Stochastic Processes and their Applications.* **126** (2016), 1433–1471.
15. A. W. Van Der Vaart and J. A. Wellner. Weak convergence. In *Weak convergence and empirical processes*, pp. 16–28. Springer, 1996.
16. N. Fournier and A. Guillin, On the rate of convergence in Wasserstein

distance of the empirical measure, *Probability Theory and Related Fields.* **162** (2015), 707–738.

17. J. Weed and F. Bach, Sharp asymptotic and finite-sample rates of convergence of empirical measures in Wasserstein distance, *Bernoulli.* **25** (2019), 2620–2648.

18. E. Stewart, R. Madden, G. Paul, and F. Taddei, Aging and death in an organism that reproduces by morphologically symmetric division, *PLoS Biol.* **3** (2005).

19. L. Robert, M. Hoffmann, N. Krell, S. Aymerich, J. Robert, and M. Doumic, Division in Escherichia coli is triggered by a size-sensing rather than a timing mechanism, *BMC Biology.* **12** (2014), 17. doi: 10.1186/1741-7007-12-17.

20. A. Amir, Cell size regulation in bacteria, *Phys. Rev. Lett.* **112** (2014), 208102.

21. S. Taheri-Araghi, S. Bradde, J. T. Sauls, N. S. Hill, P. A. Levin, J. Paulsson, M. Vergassola, and S. Jun, Cell-size control and homeostasis in bacteria, *Current Biology.* **11679** (2015), 1–7.

22. F. Si, G. Le Treut, J. T. Sauls, S. Vadia, P. A. Levin, and S. Jun, Mechanistic origin of cell-size control and homeostasis in bacteria, *Current Biology.* **29** (2019), 1760–1770.

23. B. Delyon, B. de Saporta, N. Krell, and L. Robert, Investigation of asymmetry in E. coli growth rate, *Case Studies in Business, Industry & Government Statistics.* **7** (2018).

24. J. Lamperti, Continuous state branching processes, *Bulletin of the American Mathematical Society.* **73** (1967), 382–386.

25. D. G. Kendall, Branching processes since 1873, *Journal of the London Mathematical Society.* **1** (1966), 385–406.

26. T. E. Harris, Branching processes, *The Annals of Mathematical Statistics.* (1948), 474–494.

27. B. A. Sevastyanov, Limit theorems for branching stochastic processes of special form, *Theory of Probability & Its Applications.* **2** (1957), 321–331.

28. P. Haccou, P. Jagers, and V. Vatutin, *Branching processes: variation, growth, and extinction of populations.* Number 5, Cambridge university press, 2005.

29. S. Méléard and V. C. Tran, Trait substitution sequence process and canonical equation for age-structured populations, *Journal of Mathematical Biology.* **58** (2009), 881–921.

30. N. Champagnat, R. Ferrière, and S. Méléard, Unifying evolutionary dynamics: from individual stochastic processes to macroscopic models, *Theoretical population biology.* **69** (2006), 297–321.

31. N. Champagnat, R. Ferrière, and S. Méléard, From individual stochastic processes to macroscopic models in adaptive evolution, *Stochastic Models.* **24** (2008), 2–44.

32. V. Bansaye, J.-F. Delmas, L. Marsalle, and V. C. Tran, Limit theorems for Markov processes indexed by continuous time Galton-Watson trees, *The Annals of Applied Probability.* **21** (2011), 2263–2314.

33. V. Bansaye, M.-E. Caballero, and S. Méléard, Scaling limits of population

and evolution processes in random environment, *Electronic Journal of Probability.* **24** (2019), 1–38.

34. B. Perthame, *Transport equations in biology.* Frontiers in Mathematics, Birkhäuser Verlag, Basel, 2007.

35. P. Michel, S. Mischler, and B. Perthame, General relative entropy inequality: an illustration on growth models, *J. Math. Pures Appl. (9).* **84** (2005), 1235–1260.

36. M. Doumic and P. Gabriel, Eigenelements of a general aggregation-fragmentation model, *Mathematical Models and Methods in Applied Sciences.* **20** (2009), 757.

37. M. Doumic, M. Hoffmann, N. Krell, and L. Robert, Statistical estimation of a growth-fragmentation model observed on a genealogical tree, *Bernoulli.* **21** (2015), 1760–1799.

38. F. Baccelli, B. Błaszczyszyn, and M. Karray, Random measures, point processes, and stochastic geometry, 2020.

39. D. J. Daley and D. Vere-Jones, *An introduction to the theory of point processes: volume I: elementary theory and methods.* Springer, 2003.

40. J. Jacod and A. Shiryaev, *Limit theorems for stochastic processes.* vol. 288, Springer Science & Business Media, 2013.

41. Tran, Viet Chi, Large population limit and time behaviour of a stochastic particle model describing an age-structured population, *ESAIM: PS.* **12** (2008), 345–386.

42. N. Fournier and S. Méléard, A microscopic probabilistic description of a locally regulated population and macroscopic approximations, *The Annals of Applied Probability.* **14** (2004), 1880–1919.

43. J. Bertoin, *Random fragmentation and coagulation processes.* vol. 102, Cambridge University Press, 2006.

44. B. Haas, Loss of mass in deterministic and random fragmentations, *Stochastic processes and their applications.* **106** (2003), 245–277.

45. W. Kermack and A. McKendrick, A contribution to the mathematical theory of epidemics, *Proc. Roy. Society of London, Series A.* **115** (1927), 700–721.

46. J. A. J. Metz and O. Diekmann, eds., *The dynamics of physiologically structured populations.* vol. 68, *Lecture Notes in Biomathematics*, Springer-Verlag, Berlin, 1986. ISBN 3-540-16786-2. Papers from the colloquium held in Amsterdam, 1983.

47. T. E. Harris, *The theory of branching processes.* vol. 6, Springer Berlin, 1963.

48. W. Feller, On the integral equation of renewal theory. In *Selected Papers I*, pp. 567–591. Springer, 2015.

49. J. L. Doob, Renewal theory from the point of view of the theory of probability, *Transactions of the American Mathematical Society.* **63** (1948), 422–438.

50. A. J. Lotka, Application of recurrent series in renewal theory, *The Annals of Mathematical Statistics.* **19** (1948), 190–206.

51. P. Gabriel, Measure solutions to the conservative renewal equation, *ESAIM: Proceedings and Surveys.* **62** (2018), 68–78.

52. P. Gwiazda and E. Wiedemann, Generalized entropy method for the renewal

equation with measure data, *Communications in Mathematical Sciences.* **15** (2017), 577–586.

53. P. Gwiazda and B. Perthame, Invariants and exponential rate of convergence to steady state in the renewal equation, *Markov Process. Related Fields.* **12** (2006), 413–424.

54. S. Mischler, B. Perthame, and L. Ryzhik, Stability in a nonlinear population maturation model, *Mathematical Models and Methods in Applied Sciences.* **12** (2002), 1751–1772.

55. V. Bansaye, B. Cloez, and P. Gabriel, Ergodic behavior of non-conservative semigroups via generalized Doeblin conditions, *Acta Applicandae Mathematicae.* (2019), 1–44.

56. S. Mischler and J. Scher, Spectral analysis of semigroups and growth-fragmentation equations, *Annales de l'Institut Henri Poincaré (C) Non Linear Analysis.* **33** (2016), 849–898.

57. W. Feller, *An introduction to probability theory and its applications. Vol. II.* John Wiley and Sons Inc., 1971.

58. P. Gwiazda and B. Perthame, Invariants and exponential rate of convergence to steady state in the renewal equation, *Markov Processes and Related Fields.* **2** (2006), 413–424.

59. V. Bansaye and V. C. Tran, Branching Feller diffusion for cell division with parasite infection, *ALEA, Lat. Am. J. Probab. Math. Stat.* **8** (2011), 95–127.

60. V. Bansaye, Proliferating parasites in dividing cells: Kimmel's branching model revisited, *Ann. Appl. Probab.* **18** (2008), 967–996.

61. B. Cloez, Limit theorems for some branching measure-valued processes, *Advances in Applied Probability.* **49** (2017), 549–580.

62. O. Diekmann, H. Heijmans, and H. Thieme, On the stability of the cell size distribution, *Journal of Mathematical Biology.* **19** (1984), 227–248.

63. O. Diekmann, M. Gyllenberg, and J. Metz, Steady-state analysis of structured population models, *Theoretical population biology.* **63** (2003), 309–338.

64. M. Escobedo, A short remark on a growth–fragmentation equation, *Comptes Rendus Mathematique.* **355** (2017), 290–295.

65. M. Doumic and B. Van Brunt, Explicit solution and fine asymptotics for a critical growth-fragmentation equation, *ESAIM: Proceedings and Surveys.* **62** (2018), 30–42.

66. T. Suebcharoen, B. Van Brunt, and G. Wake, Asymmetric cell division in a size-structured growth model, *Differential and Integral Equations.* **24** (2011), 787–799.

67. P. Michel, Existence of a solution to the cell division eigenproblem, *Math. Models Methods Appl. Sci.* **16** (2006), 1125–1153.

68. P. Michel, S. Mischler, and B. Perthame, General entropy equations for structured population models and scattering, *C. R. Math. Acad. Sci. Paris.* **338** (2004), 697–702.

69. M. J. Cáceres, J. A. Cañizo, and S. Mischler, Rate of convergence to an asymptotic profile for the self-similar fragmentation and growth-fragmentation equations, *Journal de mathématiques pures et appliquées.* **96** (2011), 334–362.

70. N. Fournier and B. Perthame, A non-expanding transport distance for some structured equations, *arXiv preprint*.
71. J. A. Cañizo, P. Gabriel, and H. Yoldaş, Spectral gap for the growth-fragmentation equation via Harris theorem, *arXiv preprint arXiv:2004.08343*. (2020).
72. E. Bernard and P. Gabriel, Asymptotic behavior of the growth-fragmentation equation with bounded fragmentation rate, *Journal of Functional Analysis*. **272** (2017), 3455–3485.
73. D. Balagué, J. Cañizo, and P. Gabriel, Fine asymptotics of profiles and relaxation to equilibrium for growth-fragmentation equations with variable drift rates, *Kinetic and related models*. **6** (2013), 219–243.
74. J. Bertoin, On a Feynman-Kac approach to growth-fragmentation semigroups and their asymptotic behaviors, *Journal of Functional Analysis*. **277** (2019), 108270.
75. J. Bertoin and A. R. Watson, The strong malthusian behavior of growth-fragmentation processes, *Annales Henri Lebesgue*. **3** (2020), 795–823.
76. B. Haas, Regularity of formation of dust in self-similar fragmentations. In *Annales de l'IHP Probabilités et statistiques*, vol. 40, pp. 411–438, 2004.
77. B. Haas, Asymptotic behavior of solutions of the fragmentation equation with shattering: an approach via self-similar Markov processes, *Annals of Applied Probability*. **20** (2010), 382–429.
78. C. Goldschmidt and B. Haas, Behavior near the extinction time in self-similar fragmentations i: the stable case, *Annales de l'I.H.P. Probabilités et statistiques*. **46** (2010), 338–368. doi: 10.1214/09-AIHP317.
79. P. Dyszewski, N. Gantert, S. G. Johnston, J. Prochno, and D. Schmid, Sharp concentration for the largest and smallest fragment in a k-regular self-similar fragmentation, *arXiv preprint arXiv:2102.08935*. (2021).
80. C. Goldschmidt, and B. Haas, Behavior near the extinction time in self-similar fragmentations ii: Finite dislocation measures, *Annals of Probability*. **44** (2016), 739–805.
81. J. Banasiak, Shattering and non-uniqueness in fragmentation models; an analytic approach, *Physica D: Nonlinear Phenomena*. **222** (2006), 63–72.
82. J. Banasiak, W. Lamb, and P. Laurençot, *Analytic Methods for Coagulation-Fragmentation Models, Volume I*. CRC Press, 2019.
83. M. Doumic and M. Escobedo, Time asymptotics for a critical case in fragmentation and growth-fragmentation equations, *Kinetic & Related Models*. **9** (2016), 251.
84. M. Escobedo, On the non existence of non negative solutions to a critical growth-fragmentation equation. In *Annales de la Faculté des sciences de Toulouse: Mathématiques*, vol. 29, pp. 177–220, 2020.
85. J. Bertoin and A. R. Watson, Probabilistic aspects of critical growth-fragmentation equations, *Advances in Applied Probability*. **48** (2016), 37–61.
86. G. Greiner and R. Nagel, Growth of cell populations via one-parameter semigroups of positive operators. In *Mathematics applied to science*, pp. 79–105. Elsevier, 1988.
87. E. Bernard, M. Doumic, and P. Gabriel, Cyclic asymptotic behaviour of a

population reproducing by fission into two equal parts, *Kinetic and Related Models.* **12** (2019), 551.

88. P. Gabriel and H. Martin, Periodic asymptotic dynamics of the measure solutions to an equal mitosis equation, *arXiv preprint arXiv:1909.08276.* (2019).

89. J. A. Carrillo, P. Gwiazda, and A. Ulikowska, Splitting-particle methods for structured population models: convergence and applications, *Mathematical Models and Methods in Applied Sciences.* **24** (2014), 2171–2197.

90. V. Bansaye and S. Méléard, *Stochastic models for structured populations.* vol. 16, Springer, 2015.

91. P. Gwiazda, T. Lorenz, and A. Marciniak-Czochra, A nonlinear structured population model: Lipschitz continuity of measure-valued solutions with respect to model ingredients, *Journal of Differential Equations.* **248** (2010), 2703–2735.

92. P. Gabriel and H. Martin, Steady distribution of the incremental model for bacteria proliferation, *arXiv preprint arXiv:1803.04950.* (2018).

93. M. Doumic, Analysis of a population model structured by the cells molecular content, *Math. Model. Nat. Phenom.* **2** (2007), 121–152.

94. H. Kang, X. Huo, and S. Ruan, Nonlinear physiologically structured population models with two internal variables, *Journal of Nonlinear Science.* **30** (2020), 2847–2884.

95. H. Banks, K. Sutton, W. Thompson, G. Bocharov, D. Roosec, T. Schenkeld, and A. Meyerhanse, Estimation of cell proliferation dynamics using cfse data, *Bull. of Math. Biol.* (2010).

96. A. B. Tsybakov, *Introduction à l'estimation non paramétrique.* vol. 41, Springer Science & Business Media, 2003.

97. E. Giné and R. Nickl, *Mathematical foundations of infinite-dimensional statistical models.* Cambridge University Press, 2021.

98. M. Hoffmann and A. Olivier, Statistical analysis for structured models on trees, *Statistical Inference for Piecewise-deterministic Markov Processes.* (2018), 1–38.

99. M. Doumic, M. Hoffmann, P. Reynaud, and V. Rivoirard, Nonparametric estimation of the division rate of a size-structured population, *SIAM J. on Numer. Anal.* **50** (2012), 925–950.

100. M. Nussbaum and S. Pereverzev, The degrees of ill-posedness in stochastic and deterministic noise models, *Preprint WIAS 509.* (1999).

101. M. Doumic, B. Perthame, and J. Zubelli, Numerical solution of an inverse problem in size-structured population dynamics, *Inverse Problems.* **25** (2009), 045008.

102. M. Doumic, M. Escobedo, and M. Tournus, Estimating the division rate and kernel in the fragmentation equation, *Annales de l'Institut Henri Poincaré (C) Non Linear Analysis.* **35** (2018), 1847–1884.

103. V. H. Hoang, T. M. Pham Ngoc, V. Rivoirard, and V. C. Tran, Nonparametric estimation of the fragmentation kernel based on a partial differential equation stationary distribution approximation, *Scandinavian Journal of Statistics.* (2020), 1–40.

104. M. Hairer and J. C. Mattingly, Yet another look at Harris ergodic theorem for Markov chains. In *Seminar on Stochastic Analysis, Random Fields and Applications VI*, pp. 109–117, 2011.

105. P. H. Baxendale, Renewal theory and computable convergence rates for geometrically ergodic Markov chains, *The Annals of Applied Probability.* **15** (2005), 700–738.

106. S. V. B. Penda, M. Hoffmann, and A. Olivier, Adaptive estimation for bifurcating Markov chains, *Bernoulli.* **23** (2017), 3598–3637.

107. H. Engl, M. Hanke, and A. Neubauer, *Regularization of inverse problems.* vol. 375, *Mathematics and its Applications*, Springer, 1996.

108. T. Bourgeron, M. Doumic, and M. Escobedo, Estimating the division rate of the growth-fragmentation equation with a self-similar kernel, *Inverse Problems.* **30** (2014), 025007.

109. G. Wahba, Practical approximate solutions to linear operator equations when the data are noisy, *SIAM journal on numerical analysis.* **14** (1977), 651–667.

110. M. Nussbaum, Asymptotic equivalence of density estimation and Gaussian white noise, *The Annals of Statistics.* (1996), 2399–2430.

111. M. Doumic, A. Olivier, and L. Robert, Estimating the division rate from indirect measurements of single cells, *Discrete & Continuous Dynamical Systems-Series B.* **25** (2020).

112. J. A. Carrillo, P. Gwiazda, K. Kropielnicka, and A. K. Marciniak-Czochra, The escalator boxcar train method for a system of age-structured equations in the space of measures, *SIAM Journal on Numerical Analysis.* **57** (2019), 1842–1874.

113. A. Ramdas, N. G. Trillos, and M. Cuturi, On Wasserstein two-sample testing and related families of nonparametric tests, *Entropy.* **19** (2017), 47.

114. A. Fermanian, C. Doucet, M. Hoffmann, L. Robert, and M. Doumic, Cell-division: a plateform to select a best-fit cell division cycle model. In progress, prototype plateform available at https://celldivision.paris.inria.fr/welcome/.

115. I. Soifer, L. Robert, and A. Amir, Single-cell analysis of growth in budding yeast and bacteria reveals a common size regulation strategy, *Current Biology.* **26** (2016), 356–361.

Collective Dynamics of Lohe Type Aggregation Models

Seung-Yeal Ha

*Department of Mathematical Sciences and Research Institute of Mathematics
Seoul National University, Seoul 08826, and
Korea Institute for Advanced Study,
Hoegiro 85, Seoul 02455, Republic of Korea
syha@snu.ac.kr*

Dohyun Kim

*School of Mathematics, Statistics and Data Science
Sungshin Women's University
Seoul 02844, Republic of Korea
dohyunkim@sungshin.ac.kr*

In this chapter, we review state-of-the-art results on the collective behaviors for Lohe type first-order aggregation models. Collective behaviors of classical and quantum many-body systems have received lots of attention from diverse scientific disciplines such as applied mathematics, control theory in engineering, nonlinear dynamics of statistical physics, etc. To model such collective dynamics, several phenomenological models were proposed in literature and their emergent dynamics were extensively studied in recent years. Among them, we present two Lohe type models: *the Lohe tensor (LT) model* and *the Schrödinger-Lohe (SL) model*, and present several sufficient conditions in unified frameworks via the Lyapunov functional approach for state diameters and dynamical systems theory approach for two-point correlation functions. We also present several numerical simulation results for the SL model.

Contents

1. Introduction

The purpose of this chapter is to continue a recent review [40] on the collective behaviors of classical and quantum synchronization (or aggregation) models. From the beginning of this century, collective behaviors of many-body systems have received a lot of attention from various scientific disciplines, e.g., synchronization and swarming behaviors in biology [13, 62, 68, 69, 72, 73], decentralized control in engineering [54, 61, 65], non-convex stochastic optimization algorithms in machine learning community [16, 17, 28, 29, 51, 52, 64], etc. Mathematical modeling of such collective behaviors was begun by several pioneers, Winfree [73], Kuramoto [53] and Vicsek [70] whose works have established milestones of collective dynamics as major research subjects in applied mathematics, control theory, statistical physics, etc. Since then, several phenomenological models have been proposed, to name a few, the Cucker-Smale model [24], the swarm sphere model [18, 20, 21, 23, 37, 59], matrix aggregation models [12, 25, 26, 30, 36, 56, 57], etc. We also refer the reader to survey articles [1, 2, 27, 40, 60, 63, 66, 71, 73] for a brief introduction to collective dynamics.

In what follows, we are mainly interested in the Lohe type aggregation models (the LT model and the SL model). The LT model is a finite-dimensional aggregation model on the space of tensors with the same rank and size, and it encompasses previously introduced first-order aggregation models, whereas the SL model is an infinite-dimensional toy model describing synchronous behaviors in a quantum regime. In fact, Lohe first focused on the similarity between classical and quantum synchronizations and proposed a merely phenomenological model to capture common properties between two emergent behaviors in different regimes. However, when we focus on the dissimilarity between classical and quantum systems, one encounters several limitations of the SL model due to its quantum nature, namely *entanglement* which is a genuine quantum feature and is not observed in the classical world. We here omit detailed descriptions on the relation between quantum synchronization and entanglement which is beyond our scope.

In this chapter, we briefly review state-of-the-art results on the collective dynamics of the aforementioned two Lohe type aggregation models from a universal platform for collective behaviors, Lyapunov functional approach and dynamical systems theory approach.

The rest of the chapter is organized as follows. In Sec. 2, we introduce two Lohe type aggregation models, namely the LT model and the SL model. In Sec. 3, we review state-of-the-art results on the emergent dynamics of the LT model for homogeneous and heterogeneous ensembles by providing several sufficient frameworks for the complete state aggregation and practical aggregation. In Sec. 4, we review parallel results for the SL model compared to the LT model in Sec. 3. Finally, Sec. 5 is devoted to a brief summary and discussion on some remaining interesting issues for a future direction.

Notation: Throughout the chapter, we will see several models. As long as there are no confusion, we use handy acronyms for such models:

LT: Lohe tensor,   SL: Schrödinger-Lohe,   LM: Lohe matrix,

LHS: Lohe hermitian sphere,   SDS: swarm double sphere,

SDM: swarm double matrix.

Moreover, we use $|\cdot|$ to denote ℓ^2-norm of vectors in $\mathbb{R}^d$ or $\mathbb{C}^d$, where $\|\cdot\|$ represents L^2-norm.

2. Preliminaries

In this section, we briefly introduce two first-order aggregation models whose emergent dynamics will be discussed in the following two sections, namely *"the Lohe tensor model"* and *"the Schrödinger-Lohe model"*, separately.

2.1. *The Lohe tensor model*

To set up the stage, we begin with basic terminologies on tensors. A complex rank-m tensor can be visualized as a multi-dimensional array of complex numbers with multi-indices. The *"rank"* (or *"order"*) of a tensor is the number of indices, say a rank-m tensor with size $d_1 \times \cdots \times d_m$ is an element of $\mathbb{C}^{d_1 \times \cdots \times d_m}$. For example, scalars, vectors and matrices correspond to rank-0, 1 and 2 tensors, respectively.

Let T be a rank-m tensor with a size $d_1 \times \cdots \times d_m$. Then, we denote $(\alpha_1, \cdots, \alpha_m)$-th component of T by $[T]_{\alpha_1 \cdots \alpha_m}$, and we also set $\overline{T}$ by the rank-m tensor whose components are simply the complex conjugate of the corresponding elements in T:

$$[\overline{T}]_{\alpha_1 \cdots \alpha_m} := \overline{[T]_{\alpha_1 \cdots \alpha_m}}, \quad 1 \le \alpha_i \le d_i, \ \ 1 \le i \le m.$$

In other words, each component of $\overline{T}$ is defined as the complex conjugate of the corresponding element of T. Let $\mathcal{T}_m(\mathbb{C}; d_1 \times \cdots \times d_m)$ be the collection of all rank-m tensors with size $d_1 \times \cdots \times d_m$. For notational simplicity, we set

$$\mathbb{C}^{d_1 \times \cdots \times d_m} := \mathcal{T}_m(\mathbb{C}; d_1 \times \cdots \times d_m).$$

Then, it is a complex vector space. Several well-known first-order aggregation models, for instance, the Kuramoto model [53], the swarm sphere model [59] and the Lohe matrix model [57] can be regarded as aggregation models on the subsets of $\mathbb{R}, \mathbb{C}^d$ and $\mathbb{C}^{d \times d}$, respectively. Furthermore, we also introduce the following handy notation: for $T \in \mathbb{C}^{d_1 \times \cdots \times d_m}$ and $A \in \mathbb{C}^{d_1 \times \cdots \times d_m \times d_1 \times \cdots \times d_m}$, we set

$$[T]_{\alpha_*} := [T]_{\alpha_1 \alpha_2 \cdots \alpha_m}, \quad [T]_{\alpha_{*0}} := [T]_{\alpha_{10} \alpha_{20} \cdots \alpha_{m0}}, \quad [T]_{\alpha_{*1}} := [T]_{\alpha_{11} \alpha_{21} \cdots \alpha_{m1}},$$

$$[T]_{\alpha_{*i_*}} := [T]_{\alpha_{1i_1} \alpha_{2i_2} \cdots \alpha_{mi_m}}, \quad [T]_{\alpha_{*(1-i_*)}} := [T]_{\alpha_{1(1-i_1)} \alpha_{2(1-i_2)} \cdots \alpha_{m(1-i_m)}},$$

$$[A]_{\alpha_* \beta_*} := [A]_{\alpha_1 \alpha_2 \cdots \alpha_m \beta_1 \beta_2 \cdots \beta_m}.$$

Moreover, we can associate inner product $\langle \cdot, \cdot \rangle_{\mathrm{F}}$, namely *"Frobenius inner product"* and its induced norm $\| \cdot \|_{\mathrm{F}}$ on $\mathbb{C}^{d_1 \times \cdots \times d_m}$: for $T, S \in \mathbb{C}^{d_1 \times \cdots \times d_m}$,

$$\langle T, S \rangle_{\mathrm{F}} := \sum_{\alpha_* \in \prod_{i=1}^m \{1, \cdots, d_i\}} [\overline{T}]_{\alpha_*} [S]_{\alpha_*}, \quad \|T\|_{\mathrm{F}}^2 := \langle T, T \rangle_{\mathrm{F}}.$$

Let A_j be a *block skew-hermitian* rank-$2m$ tensor with size $(d_1 \times \cdots \times d_m) \times (d_1 \times \cdots \times d_m)$ such that

$$[\overline{A}_j]_{\alpha_{*0} \alpha_{*1}} = -[A_j]_{\alpha_{*1} \alpha_{*0}}.$$

In other words, if two blocks with the first m-indices are interchanged with the rest m-indices, then its sign is changed.

Now, we are ready to introduce the LT model on the finite ensemble $\{T_j\}_{j=1}^N \subset \mathbb{C}^{d_1 \times \cdots \times d_m}$:

$$
\begin{aligned}
[\dot{T}_j]_{\alpha_{*0}} &= [A_j]_{\alpha_{*0}\alpha_{*1}} [T_j]_{\alpha_{*1}} \\
&\quad + \sum_{i_* \in \{0,1\}^m} \kappa_{i_*} \Big([T_c]_{\alpha_{*i_*}} [\bar{T}_i]_{\alpha_{*1}} [T_i]_{\alpha_{*(1-i_*)}} - [T_i]_{\alpha_{*i_*}} [\bar{T}_c]_{\alpha_{*1}} [T_i]_{\alpha_{*(1-i_*)}} \Big),
\end{aligned}
$$

(2.1)

where κ_{i_*}'s are coupling strengths, $T_c := \frac{1}{N} \sum_{k=1}^N T_k$ is the f average of $\{T_j\}_{j1}^N$, and we used the Einstein summation convention for repeated indices in the R.H.S. of (2.1). Although the LT model (2.1) looks so complicated, interaction terms inside the parenthesis of (2.1) are designed to include interactions for the Kuramoto model, the swarm sphere model and the Lohe matrix model, and they certainly have "gain term−loss term" structure and are cubic in nature. These careful designs of interactions lead to the existence of a constant of motion.

Proposition 2.1: [45] *Let $\{T_j\}$ be a solution to* (2.1). *Then, for each* $j = 1, \cdots, N$, $\|T_j\|_{\mathrm{F}}$ *is a first integral for* (2.1):

$$
\|T_j(t)\|_{\mathrm{F}} = \|T_j(0)\|_{\mathrm{F}}, \quad t \geq 0.
$$

2.2. The Schrödinger-Lohe model

Consider a network consisting of N nodes denoted by $1, \cdots, N$, and we assume that the network topology is registered by an adjacent weighted matrix (a_{ij}).

For each $j = 1, \cdots, N$, let $\psi_j = \psi_j(x, t)$ be the wave-function of the quantum subsystem lying on the j-th node. Then, we assume that the spatial-temporal dynamics of ψ_j is governed by the Cauchy problem to the SL model: for $(x, t) \in \mathbb{R}^d \times \mathbb{R}_+$,

$$
\begin{cases}
i\partial_t \psi_j = -\dfrac{1}{2} \Delta \psi_j + V_j \psi_j + \dfrac{i\kappa}{2N} \sum_{k=1}^N a_{jk} \left(\psi_k - \dfrac{\langle \psi_j, \psi_k \rangle}{\langle \psi_j, \psi_j \rangle} \psi_j \right), \\
\psi_j(x, 0) = \psi_j^0(x), \quad \|\psi_j^0\| = 1, \quad j = 1, \cdots, N,
\end{cases}
$$

(2.2)

where κ is a coupling strength, and we have taken mass and normalized Planck's constant to be unity for the simplicity of presentation. Like the classical Schrödinger equation, system (2.2) satisfies L^2-conservation.

Proposition 2.2: *Let $\{\psi_j\}$ be a solution to (2.2). Then, $\|\psi_j(t)\|$ is a first integral:*

$$\|\psi_j(t)\| = \|\psi_j^0\|, \quad t \geq 0, \quad 1 \leq j \leq N.$$

Then, L^2-conservation and standard energy estimates yield a global existence of unique weak solutions to (2.2).

Theorem 2.3: [4, 47] *Suppose initial data satisfy $\psi_j^0 \in L^2(\mathbb{R}^d)$ for each $j = 1, \cdots, N$. Then, the Cauchy problem (2.2) admits a unique global-in-time weak solution satisfying*

$$\psi_j \in C([0, \infty); L^2(\mathbb{R}^d)), \quad j = 1, \cdots, N.$$

Moreover, if we assume $\psi_j^0 \in H^1(\mathbb{R}^d)$, then the corresponding global weak solution satisfies $\psi_j \in C([0, \infty); H^1(\mathbb{R}^d))$.

Before we close this section, we show that the SL model (2.2) reduces to the Kuramoto model as a special case. For this, we assume

$$V_j(x) =: \nu_j : \text{constant}, \qquad \psi_j(x, t) =: e^{-i\theta_j(t)}, \quad (x, t) \in \mathbb{R}^d \times \mathbb{R}_+.$$

We substitute the above ansatz into the SL model (2.2) to get

$$\dot{\theta}_j \psi_j = \nu_j \psi_j + \frac{i\kappa}{N} \sum_{k=1}^{N} a_{jk} \left(\psi_k - e^{-i(\theta_j - \theta_k)} \psi_j \right).$$

Then, we multiply $\overline{\psi}_j$ to the above relation, use $|\psi_j(t)|^2 = 1$, and compare the real part of the resulting relation to obtain the Kuramoto model [53]:

$$\dot{\theta}_j = \nu_j + \frac{\bar{\kappa}}{N} \sum_{k=1}^{N} a_{jk} \sin(\theta_k - \theta_j), \quad \bar{\kappa} := 2\kappa.$$

In the following two sections, we review emergent dynamics of the LT model and the SL model, separately. Most presented results in the following section will be provided without detailed proofs, but we may discuss brief ideas or key ingredients to give a feeling to see how proofs go.

3. The Lohe tensor model

In this section, we review the emergent dynamics of the LT model and two explicit low-rank LT models which can be related to the swarm sphere model and the Lohe matrix model.

3.1. *Emergent dynamics*

In this subsection, we review emergent dynamics of the Lohe tensor model (2.1). First, we recall two concepts of aggregations (or synchronizations) as follows.

Definition 3.1: [41, 43] Let $\{T_j\}$ be a finite ensemble of rank-m tensors whose dynamics is governed by (2.1).

(1) The ensemble exhibits complete state aggregation (synchronization) if relative states tend to zero asymptotically:

$$\lim_{t \to \infty} \max_{1 \leq i,j \leq N} \|T_i(t) - T_j(t)\|_{\mathrm{F}} = 0.$$

(2) The ensemble exhibits practical aggregation (synchronization) if magnitudes of relative states can be controlled by the principal coupling strength $\kappa_{0\cdots 0}$ as follows:

$$\lim_{\kappa_{i_*} \to \infty} \limsup_{t \to \infty} \max_{1 \leq i,j \leq N} \|T_i(t) - T_j(t)\|_{\mathrm{F}} = 0,$$

for some $i_* \in \{0,1\}^m$.

Remark 3.2: The jargon *"synchronization"* is often used in control theory and physics communities instead of aggregation. In fact, synchronization represents an adjustment of rhythms in oscillatory systems. In contrast, our systems under consideration might not be oscillatory. Thus, the authors feel more comfortable to use aggregation instead of synchronization.

For a given state ensemble $\{T_j\}$ and free flow ensemble $\{A_j\}$, we define diameters for both ensembles:

$$\mathcal{D}(T) := \max_{1 \leq i,j \leq N} \|T_i - T_j\|_{\mathrm{F}}, \quad \mathcal{D}(A) := \max_{1 \leq i,j \leq N} \|A_i - A_j\|_{\mathrm{F}}.$$

Note that $\mathcal{D}(T)$ is time-varying and Lipschitz continuous. Thus, it is differentiable a.e. and $\mathcal{D}(A)$ is constant, since A_j is a constant tensor.

Let $\{T_j\}$ be a solution to (2.1) with $\|T_j\|_{\mathrm{F}} = 1$. Then, after tedious and delicate analysis, one can derive a differential inequality for $\mathcal{D}(T)$ (see Proposition 4.2 in Ref. [45]): for a.e. $t > 0$,

$$\left| \frac{\mathrm{d}}{\mathrm{d}t} \mathcal{D}(T) + \kappa_0 \mathcal{D}(T) \right| \leq 2\kappa_0 \mathcal{D}(T)^2 + 2\hat{\kappa}_0 \|T_c^0\|_{\mathrm{F}} \mathcal{D}(T) + \mathcal{D}(A), \tag{3.1}$$

where T_c^0, κ_0 and $\hat{\kappa}_0$ are defined as follows:

$$T_c^0 := \frac{1}{N} \sum_{j=1}^{N} T_j^0, \qquad \kappa_0 := \kappa_{0\cdots0}, \qquad \hat{\kappa}_0 := \sum_{i_* \neq (0,\cdots,0)} \kappa_{i_*}.$$

Depending on whether the ensemble $\{T_j\}$ has the same free flows or heterogeneous free flows, we have the following two cases:

$$\mathcal{D}(A) = 0 : \text{homogeneous ensemble}, \quad \mathcal{D}(A) > 0 : \text{heterogeneous ensemble}.$$

Then, the emergent dynamics of (2.1) can be summarized as follows.

Theorem 3.3: [45] *The following assertions hold.*

(1) *(Emergence of complete state aggregation): Suppose system parameters and initial data $\{T_j^0\}$ satisfy*

$$\mathcal{D}(A) = 0, \quad \|T_j^0\|_F = 1, \quad j = 1, \cdots, N, \quad \kappa_0 > 0,$$

$$\hat{\kappa}_0 < \frac{\kappa_0}{2\|T_c^0\|_F}, \quad 0 < \mathcal{D}(T^0) < \frac{\kappa_0 - 2\hat{\kappa}_0\|T_c^0\|_F}{2\kappa_0}, \tag{3.2}$$

and let $\{T_j\}$ be a global solution to (2.1). Then, there exist positive constants C_0 and C_1 depending on κ_{i_*} and $\{T_j^0\}$ such that

$$C_0 e^{-(\kappa_0 + 2\hat{\kappa}_0\|T_c^0\|_F)t} \leq \mathcal{D}(T(t)) \leq C_1 e^{-(\kappa_0 - 2\hat{\kappa}_0\|T_c^0\|_F)t}, \quad t \geq 0.$$

(2) *(Emergence of practical aggregation): Suppose system parameters and initial data satisfy*

$$\|T_j^0\|_F = 1, \quad j = 1, \cdots, N, \quad \kappa_0 > 0,$$

$$0 \leq \mathcal{D}(T^0) \leq \eta_2, \quad 0 < \mathcal{D}(A) < \frac{|\kappa_0 - 2\hat{\kappa}_0\|T_c^0\|_F|^2}{8\kappa_0}, \tag{3.3}$$

where η_2 appearing in (3.3) is the largest positive root of the following quadratic equation:

$$-2\kappa_0 x^2 + (\kappa_0 - 2\hat{\kappa}_0\|T_c^0\|_F)x = \mathcal{D}(A).$$

Let $\{T_j\}$ be a global solution to system (2.1). Then, practical aggregation emerges asymptotically:

$$\lim_{\kappa_0 \to \infty} \limsup_{t \to \infty} \mathcal{D}(T(t)) = 0. \tag{3.4}$$

Proof: For a detailed proof, we refer the reader to Ref. [41]. Below, we instead provide a brief sketch on the key ingredient of proofs.

(i) Suppose that $\mathcal{D}(A) = 0$.

- Case A (Upper bound estimate): It follows from (3.1) that

$$\frac{\mathrm{d}}{\mathrm{d}t}\mathcal{D}(T) \leq \mathcal{D}(T)\Big[2\kappa_0\mathcal{D}(T) - (\kappa_0 - 2\hat{\kappa}_0\|T_c^0\|_\mathrm{F})\Big], \quad \text{a.e. } t > 0. \tag{3.5}$$

Under the assumptions (3.2), coefficients appearing in the R.H.S. of (3.5) satisfy

$$2\kappa_0 > 0, \quad \kappa_0 - 2\hat{\kappa}_0\|T_c^0\|_\mathrm{F} > 0.$$

Now, we directly solve the differential inequality (3.5) to derive desired upper bound estimates.

- Case B (Lower bound estimate): Again, it follows from (3.1) that

$$\frac{\mathrm{d}}{\mathrm{d}t}\mathcal{D}(T) \geq \mathcal{D}(T)\Big[2\kappa_0\mathcal{D}(T) - (\kappa_0 + 2\hat{\kappa}_0\|T_c^0\|_\mathrm{F})\Big], \quad \text{a.e. } t > 0. \tag{3.6}$$

Similar to Case A, we integrate (3.6) to find the desired lower bound estimate.

(ii) Suppose that $\mathcal{D}(A) > 0$. Then, it follows from (3.1) that

$$\frac{\mathrm{d}}{\mathrm{d}t}\mathcal{D}(T) \leq 2\kappa_0\mathcal{D}(T)^2 - (\kappa_0 - 2\hat{\kappa}_0\|T_c^0\|_\mathrm{F})\mathcal{D}(T) + \mathcal{D}(A), \quad \text{a.e. } t > 0. \tag{3.7}$$

In order to use a comparison principle, we introduce a quadratic function f defined by

$$f(x) := -2\kappa_0 x^2 + (\kappa_0 - 2\hat{\kappa}_0\|T_c^0\|_\mathrm{F})x. \tag{3.8}$$

Since we are interested in the regime $\kappa_0 \to \infty$, the term $\kappa_0 - 2\hat{\kappa}_0\|T_c^0\|$ will be positive. Thus, it follows from (3.7) and (3.8) that

$$\frac{\mathrm{d}}{\mathrm{d}t}\mathcal{D}(T) \leq \mathcal{D}(A) - f(\mathcal{D}(T)), \quad \text{a.e. } t > 0.$$

Considering the geometry of the graph of f, one can see that the quadratic equation $f(x) = \mathcal{D}(A)$ has two positive roots η_1 and η_2 satisfying

$$0 < \eta_1 < \frac{\kappa_0 - 2\hat{\kappa}_0\|T_c^0\|_\mathrm{F}}{4\kappa_0} < \eta_2 < \frac{\kappa_0 - 2\hat{\kappa}_0\|T_c^0\|_\mathrm{F}}{2\kappa_0}.$$

Moreover, one can claim: (see Lemma 5.3 in Ref. [45])

(1) $\frac{\mathrm{d}}{\mathrm{d}t}\mathcal{D}(T(t)) \leq 0$ almost every $t > 0$ when $\mathcal{D}(T(t)) \in [\eta_1, \eta_2]$.

(2) $\mathcal{S}(\eta_2) := \{\mathcal{D}(T(t)) < \eta_2\}$ is a positively invariant set for the LT flow generated by (2.1).

(3) There exist $t_e \geq 0$ such that

$$\mathcal{D}(T(t)) < \eta_1, \quad t \geq t_e.$$

In fact, the smaller positive root η_1 can be calculated explicitly:

$$\eta_1 = \frac{\mathcal{D}(A)}{\kappa_0 - 2\hat{\kappa}_0 \|T_c^0\|_F} \left(\frac{2}{1 + \sqrt{1 - \frac{8\kappa_0 \mathcal{D}(A)}{\kappa_0 - 2\hat{\kappa}_0\|T_c^0\|_F}}} \right).$$

Then, it follows from the claim above that

$$\limsup_{t\to\infty} \mathcal{D}(T(t)) \leq \eta_1 = \frac{\mathcal{D}(A)}{\kappa_0 - 2\hat{\kappa}_0\|T_c^0\|_F} \frac{2}{1 + \sqrt{\frac{1 - 8\kappa_0\mathcal{D}(A)}{\kappa_0 - 2\hat{\kappa}_0\|T_c^0\|_F}}}.$$

By letting $\kappa_0 \to \infty$, one has the desired estimate (3.4).   □

3.2. Low-rank LT models

In the previous subsection, we have reviewed the emergent dynamics of the LT model in a general setting. In this subsection, we study two low-rank LT models which can be derived from the LT model on $\mathbb{C}^{d_1 \times d_2}$ and $\mathbb{C}^d$, respectively.

3.2.1. The generalized Lohe matrix model

In this part, we first derive a matrix aggregation model on $\mathbb{C}^{d_1 \times d_2}$ that can be reduced from the LT model. In the case of a square matrix $d_1 = d_2$, the Lohe matrix model serves as a first-order aggregation model on the subset of $\mathbb{C}^{d_1 \times d_2}$ which is a unitary group. Thus, the LT model can provide an aggregation model on the space of non-square matrices. More precisely, consider the Lohe tensor model (2.1) with $m = 2$:

$$\dot{T}_j = A_j T_j + \kappa_{00}(\mathrm{tr}(T_j^\dagger T_j)T_c - \mathrm{tr}(T_c^\dagger T_j)T_j) + +\kappa_{11}\mathrm{tr}(T_j^\dagger T_c - T_c^\dagger T_j)T_j$$
$$+ \kappa_{10}(T_j T_j^\dagger T_c - T_j T_c^\dagger T_j) + \kappa_{01}(T_c T_j^\dagger T_j - T_j T_c^\dagger T_j),$$
$$(3.9)$$

where $T_j^\dagger$ is a hermitian conjugate of T_j and the free flow term $A_j T_j$ is defined as a rank-2 tensor via tensor contraction between a rank-4 tensor and a rank-2 tensor:

$$[T_j]_{\alpha\beta}^\dagger = [\overline{T}_j]_{\beta\alpha}, \quad [A_j T_j]_{\alpha\beta} = [A_j]_{\alpha\beta\gamma\delta}[T_j]_{\gamma\delta}.$$

For simplicity, we set

$$\kappa_{00} = \kappa_{11} = 0, \quad \kappa_1 := \kappa_{01}, \quad \kappa_2 := \kappa_{10}.$$

Then, system (3.9) reduces to the following simplified model, namely *"the generalized Lohe matrix model"* [42]:

$$\begin{cases} \dot{T}_j = A_j T_j + \kappa_1(T_c T_j^\dagger T_j - T_j T_c^\dagger T_j) + \kappa_2(T_j T_j^\dagger T_c - T_j T_c^\dagger T_j), & t > 0, \\ T_j(0) = T_j^0 \in \mathbb{C}^{d_1 \times d_2}, & j = 1, \cdots, N, \end{cases}$$

$$(3.10)$$

where κ_1 and κ_2 are non-negative coupling strengths, and we have used Einstein summation convention.

Next, we define a functional measuring deviations from the centroid of configuration for (3.10):

$$\mathcal{V}[T(t)] := \frac{1}{N} \sum_{k=1}^N \|T_k(t) - T_c(t)\|_{\mathrm{F}}^2 = 1 - \|T_c(t)\|_{\mathrm{F}}^2, \quad t \geq 0.$$

Then, we show that $\mathcal{V}[T]$ converges to a non-negative constant $\mathcal{V}_\infty$.

Theorem 3.4: [42] *Let $\{T_j\}$ be a global solution to (3.10) with $\|T_j^0\|_{\mathrm{F}} = 1$. Then, the following assertions hold.*

(1) There exists a non-negative constant $\mathcal{V}_\infty$ such that

$$\lim_{t \to \infty} \mathcal{V}[T] = \mathcal{V}_\infty.$$

(2) The orbital derivative of $\mathcal{V}(T)$ tends to zero asymptotically:

$$\lim_{t \to \infty} \frac{\mathrm{d}}{\mathrm{d}t} \mathcal{V}[T(t)] = 0 \quad \text{and}$$

$$\lim_{t \to \infty} \left(\|T_j T_c^\dagger - T_c T_j^\dagger\|_{\mathrm{F}} + \|T_j^\dagger T_c - T_c^\dagger T_j\|_{\mathrm{F}} \right) = 0.$$

In what follows, we consider *"the reduced Lohe matrix model"* which corresponds to the generalized Lohe matrix model (3.10) with $\kappa_2 = 0$:

$$\begin{cases} \dot{T}_j = A_j T_j + \kappa_1(T_c T_j^\dagger T_j - T_j T_c^\dagger T_j), & t > 0, \\ T_j(0) = T_j^0, & j = 1, \cdots, N, \end{cases} \qquad (3.11)$$

subject to the initial conditions:

$$T_i^{0,\dagger} T_i^0 = T_j^{0,\dagger} T_j^0, \quad 1 \leq i, j \leq N.$$

Let $\{T_j\}$ be a solution to (3.11) with a specific natural frequency tensor A_j:

$$[A_j]_{\alpha\beta\gamma\delta} := [B_j]_{\alpha\gamma} \delta_{\beta\delta},$$

where B_j is a rank-2 tensor. Then, by singular value decomposition of $T_j(t)$, one has

$$T_j(t) = U_j(t)\Sigma_j V_j^\dagger.$$

Here, Σ_j and V_j are time-independent constant matrices, whereas $U_j = U_j(t)$ is a time-dependent unitary matrix satisfying

$$\begin{cases} \dot{U}_j = B_j U_j + \kappa_1 (U_c D - U_j D^\dagger U_c^\dagger U_j), & t > 0, \\ U_j(0) = U_j^0, & j = 1, \cdots, N, \end{cases} \tag{3.12}$$

where D is a diagonal matrix and $B_j U_j$ is a usual matrix multiplication between B_j and U_j. Moreover, if complete state aggregation occurs for (3.11), then it also occurs for (3.12), and vice versa. By algebraic manipulation, one can find a differential inequality for the diameter of $\{U_j\}$:

$$\mathcal{D}(U) := \max_{1 \leq i,j \leq N} \|U_i - U_j\|_F, \quad \mathcal{D}(B) := \max_{1 \leq i,j \leq N} \|B_i - B_j\|_F.$$

For notational simplicity, we denote

$$D := \mathrm{diag}(\lambda_1^2, \cdots, \lambda_{d_1}^2), \quad \langle \lambda^2 \rangle := \frac{1}{d_1}(\lambda_1^2 + \lambda_2^2 + \cdots + \lambda_{d_1}^2),$$

$$\Delta(\lambda^2) := \max_{1 \leq k \leq d_1} |\lambda_k^2 - \langle \lambda^2 \rangle|, \quad \mathcal{A} := \langle \lambda^2 \rangle + \Delta(\lambda^2), \quad \mathcal{B} := \langle \lambda^2 \rangle - \Delta(\lambda^2).$$

Now, we are ready to provide the emergent dynamics of (3.12) as follows.

Theorem 3.5: [42] *The following assertions hold.*

(1) *(Complete state aggregation): Suppose system parameters and initial data satisfy*

$$\mathcal{D}(B) = 0, \quad \mathcal{A} > 0, \quad \mathcal{B} > 0, \quad \kappa_1 > 0, \quad \mathcal{D}(U^0) \leq \sqrt{\frac{2\mathcal{B}}{\mathcal{A}}}.$$

Then for any solution $\{U_j\}$ to (3.12), we have

$$\lim_{t \to \infty} \mathcal{D}(U) = 0.$$

Moreover, the convergence rate is exponential.

(2) *(Practical aggregation): Suppose system parameters and initial data satisfy*

$$\mathcal{A} > 0, \quad \mathcal{B} > 0, \quad \kappa_1 > \mathcal{D}(B) \cdot \sqrt{\frac{27\mathcal{A}}{32\mathcal{B}^3}} > 0, \quad \mathcal{D}(U^0) < \alpha_2,$$

where α_2 is a largest positive root of $g(x) = \mathcal{A}x^3 - 2\mathcal{B}x + \frac{\mathcal{D}(B)}{\kappa_1} = 0$, and let $\{U_j\}$ be a solution to (3.1). Then, one has practical aggregation:

$$\lim_{\kappa_1 \to \infty} \limsup_{t \to \infty} \mathcal{D}(U) = 0.$$

Proof: The first statement for the complete state aggregation is based on the following differential inequalities:

$$-2\kappa_1 \mathcal{A}\mathcal{D}(U) + \kappa_1 \mathcal{A}\mathcal{D}(U)^3 \leq \frac{d}{dt}\mathcal{D}(U) \leq -2\kappa_1 \mathcal{B}\mathcal{D}(U) + \kappa_1 \mathcal{A}\mathcal{D}(U)^3,$$

where $\mathcal{A}$ and $\mathcal{B}$ are constants determined by the diagonal matrix D. This implies the desired upper and lower bound estimates for $\mathcal{D}(U)$. In contrast, the proof of the second statement will be done by similar arguments as in Theorem 3.3. For details, we refer the reader to Ref. [42]. □

3.2.2. *The Lohe hermitian sphere model*

In this part, we consider a reduction of the LT model on the hermitian sphere $\{z \in \mathbb{C}^d : |z| = 1\}$. In this case, the LT model reduces to

$$\dot{z}_j = \Omega_j z_j + \kappa_0 (z_c \langle z_j, z_j \rangle - z_j \langle z_c . z_j \rangle) + \kappa_1 (\langle z_j, z_c \rangle - \langle z_c z_j \rangle) z_j, \quad (3.13)$$

where $\langle z, w \rangle$ and Ω_j are standard inner product in $\mathbb{C}^d$ and a skew-hermitian $d \times d$ matrix satisfying

$$\langle z, w \rangle = [\bar{z}]_\alpha [w]_\alpha, \quad \Omega_j^\dagger = -\Omega_j.$$

Here we used the Einstein summation convention. We refer the reader to Refs. [14, 15, 32] for the emergent dynamics of (3.13).

Note that for a real-valued rank-1 tensor $z_j \in \mathbb{R}^d$, the coupling terms involving κ_1 become identically zero thanks to the symmetry of inner product in $\mathbb{R}^d$. Hence, we can recover the swarm sphere model on $\mathbb{S}^{d-1}$:

$$\dot{x}_j = \Omega_j x_j + \kappa_0 \left(\langle x_j, x_j \rangle x_c - \langle x_c, x_j \rangle x_j \right). \quad (3.14)$$

The emergent dynamics of (3.14) has been extensively studied in a series of papers [18, 20–23, 50, 56–58, 67, 74]. In what follows, to investigate the nonlinear effect on the collective behaviors of (3.13) due to two coupling terms, we consider for a while the following two special cases:

(i) $\Omega_j = \Omega$, $\kappa_1 = 0$;   (ii) $\Omega_j = \Omega$, $\kappa_0 = 0$, $j = 1, \cdots, N$.

Then, the corresponding models for each case read as follows:

$$(i) \quad \dot{z}_j = \Omega z_j + \kappa_0(\langle z_j, z_j \rangle z_c - \langle z_c, z_j \rangle z_j).$$
$$(ii) \quad \dot{z}_j = \Omega z_j + \kappa_1(\langle z_j, z_c \rangle - \langle z_c, z_j \rangle)z_j. \qquad (3.15)$$

From the models in (3.15), we define a functional for $\{z_j\}$:

$$\mathcal{D}(Z) := \max_{1 \le i,j \le N} |1 - \langle z_i, z_j \rangle|.$$

In the following proposition, we summarize results on the emergent dynamics of the models in (3.15) without proofs.

Proposition 3.6: [44] *The following assertions hold.*

(1) Suppose system parameters and initial data satisfy

$$\kappa_0 > 0, \quad |z_j^0| = 1, \quad \max_{i \ne j} |1 - \langle z_i^0, z_j^0 \rangle| < \frac{1}{2},$$

and let $\{z_j\}$ be a global solution to $(3.15)_1$ with the initial data $\{z_j^0\}$. Then, there exists a positive constant Λ depending on the initial data such that

$$\mathcal{D}(Z(t)) \le \mathcal{D}(Z^0)e^{-\kappa_0 \Lambda t}, \quad t \ge 0.$$

(2) Suppose system parameters and initial data satisfy

$$\kappa_1 > 0, \quad |z_j^0| = 1,$$

and let $\{z_j\}$ be a global solution to $(3.15)_2$ with the initial data $\{z_j^0\}$. Then, there exist a time-dependent phase function $\theta_j = \theta_j(t)$ such that

$$z_j(t) = e^{i\theta_j(t)} z_j^0, \quad j = 1, \cdots, N,$$

and θ_j is a solution to the Kuramoto-type model with frustrations:

$$\begin{cases} \dot{\theta}_j = \dfrac{2\kappa_1}{N} \displaystyle\sum_{k=1}^{N} R_{jk}^0 \sin(\theta_k - \theta_j + \alpha_{jk}^0), \quad t > 0, \\ \theta_j(0) = 0, \quad j = 1, \cdots, N, \end{cases} \qquad (3.16)$$

where R_{jk}^0 and α_{jk}^0 are determined by initial data:

$$\langle z_j^0, z_k^0 \rangle = R_{jk}^0 e^{i\alpha_{jk}^0}.$$

Remark 3.7: (i) In Proposition 3.6(2), R_{jk}^0 and α_{jk}^0 satisfy symmetry and anti-symmetry properties, respectively:

$$R_{jk}^0 = R_{kj}^0, \qquad \alpha_{jk}^0 = -\alpha_{kj}^0, \qquad j, k = 1, \cdots, N.$$

(ii) System (3.16) can be rewritten as a gradient flow with the following potential V:

$$\dot{\Theta} = -\nabla_\Theta V[\Theta], \quad t > 0, \quad V[\Theta] := \frac{\kappa_1}{N} \sum_{i,j=1}^N R_{ij}^0 \left(1 - \cos(\theta_i - \theta_j + \alpha_{ji}^0) \right).$$

We now return to the full model (3.13) with the same free flows. Note that the term involving κ_0 corresponds to the swarm sphere model and the term with κ_1 describes the complex nature of underlying phase space. For an ensemble $\{z_j\}$, we define the norm of the centroid:

$$\rho(t) := \left| \frac{1}{N} \sum_{j=1}^N z_j(t) \right|.$$

Theorem 3.8: [44] *Suppose system parameters and initial data satisfy*

$$\Omega_j = \Omega, \quad j = 1, \cdots, N, \quad 0 < \kappa_1 < \frac{1}{4}\kappa_0, \quad \rho^0 > \frac{N-2}{N}, \tag{3.17}$$

and let $\{z_j\}$ be a solution to (3.13). Then for each $i, j = 1, \cdots, N$, two-point correlation function $\langle z_i, z_j \rangle$ converges to 1 exponentially fast, i.e., complete state aggregation emerges asymptotically.

Proof: We give a brief sketch for a proof. Details can be found in Theorem 4.1 [44]. First, note that

$$|z_i - z_j|^2 = |z_i|^2 + |z_j|^2 - 2\mathrm{Re}\langle z_i, z_j \rangle = 2(1 - \mathrm{Re}\langle z_i, z_j \rangle) \le 2|1 - \langle z_i, z_j \rangle|.$$

Thus, once we can show that $|1 - \langle z_i, z_j \rangle|$ tends to zero exponentially fast, then it directly follows that $\mathcal{D}(Z)$ tends to zero exponentially fast. Thus, we introduce a functional

$$\mathcal{L}(Z) := \max_{1 \le i,j \le N} |1 - \langle z_i, z_j \rangle|^2.$$

By detailed and straightforward calculation, one can derive differential inequality for $\mathcal{L}(Z)$:

$$\frac{\mathrm{d}}{\mathrm{d}t}\mathcal{L}(Z) \le -\kappa_0 \mathcal{L}(Z) \left(\mathrm{Re}(\langle z_{i_0} + z_{j_0}, z_c \rangle) - \frac{4\kappa_1}{\kappa_0} \right),$$

where i_0 and j_0 are extremal indices such that

$$\mathcal{L}(Z) =: |1 - \langle z_{i_0}, z_{j_0} \rangle|^2.$$

On the other hand, one can show that under the assumption (3.17) on ρ^0, the quantity $\langle z_i, z_c \rangle$ tends to 1 asymptotically. Again, by the assumption on the coupling strengths, there exist positive constants T and ε such that

$$\mathrm{Re}(\langle z(t)_{i_0} + z(t)_{j_0}, z_c \rangle) - \frac{4\kappa_1}{\kappa_0} > \varepsilon, \quad t > T.$$

This yields

$$\frac{d}{dt}\mathcal{L}(Z) \le -\kappa_0 \varepsilon \mathcal{L}(Z), \quad t > T.$$

Hence, one gets the exponential decay of $\mathcal{L}(Z)$.   □

Before we close this subsection, we discuss the swarm double sphere (SDS) model on $\mathbb{S}^{d_1-1} \times \mathbb{S}^{d_2-1}$ which was recently introduced by Lohe [55]:

$$\begin{cases} \dot{u}_i = \Omega_i u_i + \dfrac{\kappa}{N} \displaystyle\sum_{j=1}^{N} \langle v_i, v_j \rangle (u_j - \langle u_i, u_j \rangle u_i), \quad t > 0, \\[2mm] \dot{v}_i = \Lambda_i v_i + \dfrac{\kappa}{N} \displaystyle\sum_{j=1}^{N} \langle u_i, u_j \rangle (v_j - \langle v_i, v_j \rangle v_i), \\[2mm] (u_i, v_i)(0) = (u_i^0, v_i^0) \in \mathbb{S}^{d_1-1} \times \mathbb{S}^{d_2-1}, \quad 1 \le i \le N, \end{cases} \tag{3.18}$$

where $\Omega_i \in \mathbb{R}^{d_1 \times d_1}$ and $\Lambda_i \in \mathbb{R}^{d_2 \times d_2}$ are real skew-symmetric matrices, respectively, and κ denotes the (uniform) non-negative coupling strength. For homogeneous zero free flows

$$\Omega_i = O_d, \quad \Lambda_i = O_d, \quad i = 1, \cdots, N,$$

system (3.18) can be represented as a gradient flow. More precisely, we set an analytical potential $\mathcal{E}_s$ as

$$\mathcal{E}_s(U, V) := 1 - \frac{1}{N^2} \sum_{i,j=1}^{N} \langle u_i, u_j \rangle \langle v_i, v_j \rangle.$$

Then, system (3.18) can recast as a gradient system on the compact state space $(\mathbb{S}^{d_1-1} \times \mathbb{S}^{d_2-1})^N$:

$$\begin{cases} \dot{u}_i = -\dfrac{N\kappa}{2} \mathbb{P}_{T_{u_i}\mathbb{S}^{d_1-1}} \Big(\nabla_{u_i} \mathcal{E}_s(U, V) \Big), \\[2mm] \dot{v}_i = -\dfrac{N\kappa}{2} \mathbb{P}_{T_{v_i}\mathbb{S}^{d_2-1}} \Big(\nabla_{v_i} \mathcal{E}_s(U, V) \Big), \end{cases} \tag{3.19}$$

where projection operators onto the tangent spaces of $\mathbb{S}^{d_1-1}$ and $\mathbb{S}^{d_2-1}$ at u_i and v_i, respectively, are defined by the following explicit formula: for $w_1 \in \mathbb{R}^{d_1}$ and $w_2 \in \mathbb{R}^{d_2}$,

$$\mathbb{P}_{T_{u_i}\mathbb{S}^{d_1-1}}(w_1) := w_1 - \langle w_1, u_i \rangle u_i, \quad \mathbb{P}_{T_{v_i}\mathbb{S}^{d_2-1}}(w_2) := w_2 - \langle w_2, v_i \rangle v_i.$$

By the standard convergence result on a gradient system with analytical potential on a compact space, one can derive the following convergence result for all initial data.

Proposition 3.9: *The following assertions hold.*

(1) *Let $\{(u_i, v_i)\}$ be a solution to (3.19). Then, there exists a constant asymptotic state $(U^\infty, V^\infty) \in (\mathbb{S}^{d_1-1})^N \times (\mathbb{S}^{d_2-1})^N$ such that*

$$\lim_{t \to \infty}(U(t), V(t)) = (U^\infty, V^\infty).$$

(2) *Suppose initial data satisfy*

$$\min_{1 \leq i,j \leq N}\langle u_i^0, u_j^0 \rangle > 0, \quad \min_{1 \leq i,j \leq N}\langle v_i^0, v_j^0 \rangle > 0, \qquad (3.20)$$

and let $\{(u_i, v_i)\}$ be a solution to system (3.19). Then, one has complete state aggregation:

$$\lim_{t \to \infty}\max_{1 \leq i,j \leq N}|u_i(t) - u_j(t)| = 0, \quad \lim_{t \to \infty}\max_{1 \leq i,j \leq N}|v_i(t) - v_j(t)| = 0.$$

Remark 3.10: 1. We give some comments on the results in Proposition 3.9. In the first statement, the asymptotic state (U^∞, V^∞) may depend on initial data. Thus, the result does not tell us whether the complete state aggregation occurs or not as it is.

2. The second result says that relative states $u_i - u_j$ and $v_i - v_j$ tend to zero asymptotically, but if we combine both results (1) and (2), we can show that the initial data satisfying (3.20) lead to complete state aggregation.

So far, we have considered sufficient conditions for the emergence of complete state aggregation and practical aggregation. However, this emergent dynamics does not tell us the solution structure of the LT model. In the following subsection, we consider a special set of solutions, namely tensor product states which can be expressed as tensor products of lower rank tensors.

3.3. *Tensor product states*

In this subsection, we review tensor product states for the LT model which can be written as a tensor product of rank-1 tensors or rank-2 tensors. In the following definition, we provide concepts of two special tensor product states.

Definition 3.11: [38,39]

(1) Let $\{T_i\}$ be a "completely separable state" if it is a solution to (2.1), and it is the tensor product of only rank-1 tensors with unit modulus: for $1 \leq i \leq N$ and $1 \leq k \leq m$,

$$T_i = u_i^1 \otimes u_i^2 \otimes \cdots \otimes u_i^m, \quad u_i^k \in \mathbb{C}^{d_k}, \quad |u_i^k| = 1,$$

where $|\cdot|$ is the standard ℓ^2-norm in $\mathbb{C}^d$.

(2) Let $\{T_i\}$ be a "quadratically separable state" if it is a solution to (2.1), and it is the tensor product of only rank-2 tensors (or matrices) with unit Frobenius norm: for $1 \leq i \leq N$ and $1 \leq k \leq m$,

$$T_i = U_i^1 \otimes U_i^2 \otimes \cdots \otimes U_i^m, \quad U_i^k \in \mathbb{C}^{d_1^k \times d_2^k}, \quad \|U_i^k\|_{\mathrm{F}} = 1,$$

where $\|\cdot\|_{\mathrm{F}}$ is the Frobenius norm induced by Frobenius inner product.

3.3.1. *Completely separable state*

To motivate our discussion, we begin with rank-2 tensors that can be decomposed into two rank-1 tensors for all time. Then, since its extension to the case of a rank-m tensor will be straightforward, it suffices to focus on rank-2 tensors, and we refer the reader to Refs. [38,39] for details. In the next proposition, we show that (3.10) and (3.18) are equivalent in the following sense.

Proposition 3.12: [39] *The following assertions hold.*

(1) *(Construction of a completely separable state): Suppose* $\{(u_i, v_i)\}$ *is a global solution to (3.18). Then, a real rank-2 tensor T_i defined by $T_i := u_i \otimes v_i$ is a completely separable state to (3.10) with a well-prepared free flow tensor A_i and coupling strengths:*

$$A_i T_i := \Omega_i T_i + T_i \Lambda_i^\top, \quad \kappa_1 = \kappa_2 =: \kappa.$$

(2) *(Propagation of complete separability): Suppose T_i is a solution to (3.10) with completely separable initial data:*

$$T_i^0 =: u_i^0 \otimes v_i^0, \quad 1 \le i, j \le N,$$

for real rank-1 tensors $u_i^0 \in \mathbb{S}^{d_1-1}$ and $v_i^0 \in \mathbb{S}^{d_2-1}$. Then, there exist two unit vectors $u_i = u_i(t)$ and $v = v_i(t)$ such that

$$T_i(t) = u_i(t) \otimes v_i(t), \quad t > 0,$$

where (u_i, v_i) is a solution to (3.18) with $(u_i, v_i)(0) = (u_i^0, v_i^0)$.

Thus, in order to investigate the emergent dynamics of some classes of solutions to (3.10), it suffices to study the dynamics of (3.18).

Theorem 3.13: [39] *Let $\{T_i = u_i \otimes v_i\}$ be a completely separable state to (3.10) with the initial data $\{u_i^0 \otimes v_i^0\}$ satisfying*

$$\min_{1 \le i,j \le N} \langle u_i^0, u_j^0 \rangle > 0 \quad \text{and} \quad \min_{1 \le i,j \le N} \langle v_i^0, v_j^0 \rangle > 0.$$

Then, we have complete state aggregation:

$$\lim_{t \to \infty} \|T_i(t) - T_j(t)\|_{\mathrm{F}} = 0, \quad 1 \le i, j \le N.$$

Proof: Note that

$$T_i - T_j = u_i \otimes v_i - u_j \otimes v_j = u_i v_i^\top - u_j v_j^\top = (u_i - u_j)v_i^\top + u_j(v_i^\top - v_j^\top).$$

This yields

$$
\begin{aligned}
\|T_i - T_j\|_{\mathrm{F}} &\le \|(u_i - u_j)v_i^\top\|_{\mathrm{F}} + \|u_j(v_i^\top - v_j^\top)\|_{\mathrm{F}} \\
&\le \|u_i - u_j\|_{\mathrm{F}} \cdot \|v_i^\top\|_{\mathrm{F}} + \|u_j\|_{\mathrm{F}} \cdot \|v_i^\top - v_j^\top\|_{\mathrm{F}} \\
&= |u_i - u_j| \cdot |v_i| + |u_j| \cdot |v_i - v_j| = |u_i - u_j| + |v_i - v_j|.
\end{aligned}
$$

By the result of Proposition 3.12, one has complete state aggregation of (3.10) under suitable conditions on initial data and coupling strengths, and the desired estimates follow.   □

Remark 3.14: Extension to rank-m tensors of the results in Theorem 3.13 can be found in Secs. 6 and 7 of Ref. [39].

3.3.2. Quadratically separable state

Similar to the previous part, we consider only a finite ensemble of rank-4 tensors that can be decomposed into a tensor product of two rank-2 tensors, and extension to rank-$2m$ tensors will be straightforward.

Now, we introduce the *swarm double matrix (SDM) model* induced from the LT model whose elements have rank-4 with a specific condition on natural frequencies B_j and C_j in the same spirit of the SDS model (3.18):

$$
\begin{cases}
\dot{U}_j = B_j U_j + \dfrac{\kappa_1}{N} \sum_{k=1}^{N} \left(\langle V_j, V_k \rangle_{\mathrm{F}}\, U_k U_j^{\dagger} U_j - \langle V_k, V_j \rangle_{\mathrm{F}}\, U_j U_k^{\dagger} U_j \right) \\
\qquad + \dfrac{\kappa_2}{N} \sum_{k=1}^{N} \left(\langle V_j, V_k \rangle_{\mathrm{F}}\, U_j U_j^{\dagger} U_k - \langle V_k, V_j \rangle_{\mathrm{F}}\, U_j U_k^{\dagger} U_j \right), \\
\dot{V}_j = C_j V_j + \dfrac{\kappa_1}{N} \sum_{k=1}^{N} \left(\langle U_j, U_k \rangle_{\mathrm{F}}\, V_k V_j^{\dagger} V_j - \langle U_k, U_j \rangle_{\mathrm{F}}\, V_j V_k^{\dagger} V_j \right) \\
\qquad + \dfrac{\kappa_2}{N} \sum_{k=1}^{N} \left(\langle U_j, U_k \rangle_{\mathrm{F}}\, V_j V_j^{\dagger} V_k - \langle U_k, U_j \rangle_{\mathrm{F}}\, V_j V_k^{\dagger} V_j \right),
\end{cases}
\tag{3.21}
$$

where $B_j \in \mathbb{C}^{d_1 \times d_2 \times d_1 \times d_2}$ and $C_j \in \mathbb{C}^{d_3 \times d_4 \times d_3 \times d_4}$ are block skew-hermitian rank-4 tensors, respectively. Similar to the SDS model in Sec. 3.2.2, the SDM model (3.21) with homogeneous zero free flows can be rewritten as a gradient system for the following analytical potential:

$$
\mathcal{E}_m(U, V) := 1 - \frac{1}{N^2} \sum_{i,j=1}^{N} \langle U_i, U_j \rangle_{\mathrm{F}} \langle V_i, V_j \rangle_{\mathrm{F}}.
$$

We refer the reader to Ref. [38] for details.

Next, we consider a special case for the SDM model:

$$
d_1 = d_2 = n, \quad d_3 = d_4 = m, \quad U_j^0 \in \mathbf{U}(n), \quad V_j^0 \in \mathbf{U}(m),
$$

where $\mathbf{U}(n)$ and $\mathbf{U}(m)$ denote $n \times n$ and $m \times m$ unitary groups, respectively. In this case, one can easily show that unitary properties of U_j and V_j are propagated along (3.21):

$$
U_j(t) \in \mathbf{U}(n), \quad V_j(t) \in \mathbf{U}(m), \quad j = 1, \cdots, N, \quad t > 0. \tag{3.22}
$$

Note that natural frequency tensors B_j and C_j are rank-4 tensors satisfying block skew-hermitian properties. In order to give a meaning of Hamiltonian, we associate two hermitian matrices, namely, $H_j \in \mathbb{C}^{n \times n}$ and $G_j \in \mathbb{C}^{m \times m}$:

$$[B_j]_{\alpha_1 \beta_1 \alpha_2 \beta_2} =: [-iH_j]_{\alpha_1 \alpha_2} \delta_{\beta_1 \beta_2}, \quad [C_j]_{\gamma_1 \delta_1 \gamma_2 \delta_2} =: [-iG_j]_{\gamma_1 \gamma_2} \delta_{\delta_1 \delta_2}. \quad (3.23)$$

Under the setting (3.22) and (3.23), system (3.21) reduces to the model on $\mathbf{U}(n) \times \mathbf{U}(m)$:

$$\begin{cases} \dot{U}_j = -iH_j U_j + \dfrac{\kappa}{N} \sum_{k=1}^{N} \left(\langle V_j, V_k \rangle_{\mathrm{F}} U_k - \langle V_k, V_j \rangle_{\mathrm{F}} U_j U_k^{\dagger} U_j \right), \\[2mm] \dot{V}_j = -iG_j V_j + \dfrac{\kappa_1}{N} \sum_{k=1}^{N} \left(\langle U_j, U_k \rangle_{\mathrm{F}} V_k - \langle U_k, U_j \rangle_{\mathrm{F}} V_j V_k^{\dagger} V_j \right), \end{cases} \quad (3.24)$$

where $H_j U_j$ and $G_j V_j$ are now usual matrix products. We recall two concepts of definitions for emergent behaviors.

Definition 3.15: [46] Let $(\mathcal{U}, \mathcal{V}) := \{U_j, V_j\}_{j=1}^{N}$ be a solution to (3.24).

(1) System (3.24) exhibits complete state aggregation if the following estimate holds:

$$\lim_{t \to \infty} \max_{1 \le i,j \le N} \left(\|U_i(t) - U_j(t)\|_{\mathrm{F}} + \|V_i(t) - V_j(t)\|_{\mathrm{F}} \right) = 0.$$

(2) System (3.24) exhibits state-locking if the following relations hold:

$$\exists \lim_{t \to \infty} U_i(t) U_j(t)^{\dagger} \quad \text{and} \quad \exists \lim_{t \to \infty} V_i(t) V_j(t)^{\dagger}.$$

For the emergent dynamics of (3.24), we introduce several functionals measuring the degree of aggregation:

$$\mathcal{L}(t) := \mathcal{D}(\mathcal{U}(t)) + \mathcal{D}(\mathcal{V}(t)) + \mathcal{S}(\mathcal{U}(t)) + \mathcal{S}(\mathcal{V}(t)),$$

$$\mathcal{D}(\mathcal{U}(t)) := \max_{1 \le i,j \le N} \|U_i(t) - U_j(t)\|_{\mathrm{F}}, \quad \mathcal{S}(\mathcal{U}(t)) := \max_{1 \le i,j \le N} |n - \langle U_i, U_j \rangle_{\mathrm{F}}(t)|,$$

$$\mathcal{D}(\mathcal{V}(t)) := \max_{1 \le i,j \le N} \|V_i(t) - V_j(t)\|_{\mathrm{F}}, \quad \mathcal{S}(\mathcal{V}(t)) := \max_{1 \le i,j \le N} |m - \langle V_i, V_j \rangle_{\mathrm{F}}(t)|.$$

$$(3.25)$$

By using the unitarity of U_i and V_i, we see

$$\|U_i - U_j\|_{\mathrm{F}}^2 = 2\mathrm{Re}(n - \langle U_i, U_j \rangle_{\mathrm{F}}), \quad \|V_i - V_j\|_{\mathrm{F}}^2 = 2\mathrm{Re}(m - \langle V_i, V_j \rangle_{\mathrm{F}}).$$

Thus, one has

$$\mathcal{D}(\mathcal{U})^2 \le 2\mathcal{S}(\mathcal{U}), \quad \mathcal{D}(\mathcal{V})^2 \le 2\mathcal{S}(\mathcal{V}).$$

From the relation above, we observe

$$\lim_{t \to \infty} \mathcal{L}(t) = 0 \quad \Longleftrightarrow \quad \text{complete state aggregation,}$$

and for given positive integers n and m, we set

$$\alpha_{n,m} := \frac{-(12n+27)+\sqrt{(12n+27)^2+48(m-4\sqrt{n})(3n+4)}}{4(3n+4)}.$$

Then, the following emergent estimate can be verified by deriving a suitable dissipative differential inequality for $\mathcal{L}$ defined in (3.25).

Theorem 3.16: [38] *Suppose system parameters and initial data satisfy*

$$H_j = O_n, \quad G_j = O_m, \quad j = 1, \cdots, N, \quad n \geq m > 4\sqrt{n}, \quad \mathcal{L}^0 < \alpha_{n,m},$$

and let $\{(U_j, V_j)\}$ be a global solution to (3.24). Then, complete state aggregation emerges asymptotically.

Proof: By tedious calculation, one can derive for a.e. $t > 0$,

$$\dot{\mathcal{L}} \leq -2\kappa(m-4\sqrt{n})\mathcal{L} + \kappa(4n+9)\mathcal{L}^2 + \kappa\left(2n+\frac{8}{3}\right)\mathcal{L}^3 =: \kappa\mathcal{L}f(\mathcal{L}), \quad (3.26)$$

where f is a quadratic polynomial defined by

$$f(s) := \left(2n+\frac{8}{3}\right)s^2 + (4n+9)s - 2(m-4\sqrt{n}).$$

By assumption on n and m, the coefficient $-2(m-4\sqrt{n}) < 0$ and $f = 0$ admit a unique positive root $\alpha_{n,m}$. Moreover, one can show that the set $\{\mathcal{L}(t) < \alpha_{n,m}\}$ is positively invariant under the flow (3.24). Thus, by the phase line analysis for (3.26), one can see that

$$\lim_{t \to \infty} \mathcal{L}(t) = 0.$$

We refer the reader to Ref. [38] for details. $\square$

Remark 3.17: For heterogeneous free flows, we can also find a sufficient framework leading to state-locking (see Definition 3.15) in terms of system parameters with a large coupling strength and initial data in Ref. [38].

Finally, we consider a special ansatz for a solution T_i with rank-4 to (2.1) as follows:

$$T_i(t) \equiv U_i(t) \otimes V_i(t), \quad t > 0,$$

where $\{(U_j, V_j)\}$ is a solution to (3.21). Parallel to Proposition 3.12, we show the propagation of the tensor product structure along (3.21).

Proposition 3.18: [38] *The following assertions hold.*

(1) *(Construction of a quadratically separable state):* Suppose $\{(U_i, V_i)\}$ *is a solution to* (3.21). *Then, a rank-4 tensor* T_i *defined by* $T_i := U_i \otimes V_i$ *is a quadratically separable state to* (2.1) *with a well-prepared free flow tensor* A_i *satisfying*

$$[A_j]_{\alpha_1 \beta_1 \gamma_1 \delta_1 \alpha_2 \beta_2 \gamma_2 \delta_2} = [B_j]_{\alpha_1 \beta_1 \alpha_2 \beta_2} \delta_{\gamma_1 \gamma_2} \delta_{\delta_1 \delta_2}$$
$$+ [C_j]_{\gamma_1 \delta_1 \gamma_2 \delta_2} \delta_{\alpha_1 \alpha_2} \delta_{\beta_1 \beta_2}. \tag{3.27}$$

(2) *(Propagation of quadratic separability):* Suppose a rank-4 tensor T_i *is a solution to* (2.1) *with* (3.27) *and quadratically separable initial data:*

$$T_i^0 =: U_i^0 \otimes V_i^0, \quad 1 \le i \le N,$$

for rank-2 tensors $U_i^0 \in \mathbb{C}^{d_1 \times d_2}$ *and* $V_i^0 \in \mathbb{C}^{d_3 \times d_4}$ *with unit Frobenius norms. Then, there exist two matrices with unit Frobenius norms* $U_i = U_i(t)$ *and* $V = V_i(t)$ *such that*

$$T_i(t) = U_i(t) \otimes V_i(t), \quad t > 0,$$

where (U_i, V_i) *is a solution to* (3.21) *with* $(U_i, V_i)(0) = (U_i^0, V_i^0)$.

As we have seen in Theorem 3.13, complete state aggregation of a quadratically separable state to the LT model will be completely determined by Proposition 3.18.

4. The Schrödinger-Lohe model

In this section, we review emergent dynamics, standing wave solutions, and numerical simulations for the SL model on a network.

4.1. *Emergent dynamics*

In this subsection, we study sufficient frameworks leading to the emergent dynamics of the SL model over a network in terms of system parameters and initial data, as we have seen in previous section. First, we recall definitions of aggregation for the SL model.

Definition 4.1: [4,22,35,47,48] Let $\Psi = \{\psi_j\}$ be a global solution to (2.2).

(1) Ψ exhibits complete state aggregation if the following estimate holds:

$$\lim_{t \to \infty} \max_{1 \le i,j \le N} \|\psi_i(t) - \psi_j(t)\| = 0.$$

(2) Ψ exhibits practical aggregation if the following estimate holds:

$$\lim_{\kappa \to \infty} \limsup_{t \to \infty} \max_{1 \le i,j \le N} \|\psi_i(t) - \psi_j(t)\| = 0.$$

(3) Ψ exhibits state-locking if the following relation holds:

$$\exists \lim_{t \to \infty} \langle \psi_i(t), \psi_j(t) \rangle.$$

As we have seen from Sec. 3, the emergent dynamics of the SL model will be different whether the corresponding linear free flows are homogeneous or heterogeneous. Precisely, we define diameter for external potentials:

$$\mathcal{D}(V) := \max_{1 \le i,j \le N} \|V_i - V_j\|_{L^\infty}.$$

Then, we have the following two cases:

$$\mathcal{D}(V) = 0 : \text{homogeneous ensemble}, \quad \mathcal{D}(V) > 0 : \text{heterogeneous ensemble}.$$

4.1.1. *All-to-all network*

In this part, we list up sufficient frameworks for the emergent dynamics of (2.2) without detailed proofs. First, we consider identical one-body potentials over all-to-all network:

$$V_i = V_j, \quad i,j = 1, \cdots, N, \quad \text{i.e.,} \quad \mathcal{D}(V) = 0 \quad \text{and} \quad a_{ik} \equiv 1.$$

For a given ensemble $\Psi = \{\psi_j\}$, we set several Lyapunov functionals measuring the degree of aggregation:

$$\mathcal{D}(\Psi) := \max_{1 \le i,j \le N} \|\psi_i - \psi_j\|, \quad h_{ij} := \langle \psi_i, \psi_j \rangle, \quad \rho := \left\| \frac{1}{N} \sum_{k=1}^{N} \psi_k \right\|,$$

where $\langle \cdot, \cdot \rangle$ and $\| \cdot \|$ are L^2-inner product and its associated L^2-norm, respectively.

Theorem 4.2: [4, 18, 22, 47, 49] *(Homogeneous ensemble) Suppose system parameters and initial data satisfy*

$$\kappa > 0, \quad a_{jk} \equiv 1, \quad \mathcal{D}(V) = 0, \quad \|\psi_j^0\| = 1, \quad 1 \le j, k \le N,$$

and let $\Psi = \{\psi_j\}$ be a global solution to (2.2) with the initial data Ψ^0. Then, the following assertions hold:

(1) *(Emergence of dichotomy): one of the following holds: either complete state aggregation:*

$$\lim_{t \to \infty} \langle \psi_i, \psi_j \rangle = 1 \quad \text{for all } i, j = 1, \cdots, N,$$

or bi-polar aggregation: there exists a single index $\ell_0 \in \{1, \cdots, N\}$ such that

$$\lim_{t\to\infty} \langle \psi_i, \psi_j \rangle = 1 \quad \text{for } i, j \neq \ell_0 \quad \text{and} \quad \lim_{t\to\infty} \langle \psi_{\ell_0}, \psi_i \rangle = -1 \quad \text{for } i \neq \ell_0.$$

(2) *If initial data satisfy $\mathcal{D}(\Psi^0) < \frac{1}{2}$, then complete state aggregation occurs exponentially fast:*

$$\mathcal{D}(\Psi(t)) \lesssim e^{-\kappa t}, \quad \text{as } t \geq 0.$$

(3) *If initial data satisfy $\mathrm{Re}\langle \psi_i^0, \psi_j^0 \rangle > 0$ for all i, j, then complete state aggregation occurs in H^1-framework as well:*

$$\lim_{t\to\infty} \|\psi_i - \psi_j\|_{H^1(\mathbb{R}^d)} = 0, \quad \text{for all } i, j = 1, \cdots, N.$$

Proof: Below, we provide some ingredients without technical details.

(i) By direct calculation, we can see that ρ satisfies

$$\frac{d\rho}{dt} = \kappa \left(\rho^2 - \frac{1}{N} \sum_{k=1}^{N} \mathrm{Re}(\langle \zeta, \psi_k \rangle^2) \right),$$

$$1 - \rho(t)^2 = \frac{1}{2N^2} \sum_{j,k=1}^{N} \|\psi_j(t) - \psi_k(t)\|^2. \tag{4.1}$$

Note that complete state aggregation emerges if and only if $\rho(t)$ converges to 1 as $t \to \infty$. By $(4.1)_1$ and boundedness of ρ, the order parameter ρ is non-decreasing in time, and hence it converges to a definite value ρ_∞. After careful analysis of ρ_∞, one can see that the desired dichotomy holds (see Ref. [48]).

(ii) By direct calculation, one can derive Gronwall's inequality for $\mathcal{D}(\Psi)$ (see Ref. [22]):

$$\dot{\mathcal{D}}(\Psi) \leq \kappa \left(-\mathcal{D}(\Psi) + 2\mathcal{D}(\Psi)^2 \right), \quad t > 0.$$

This yields

$$\mathcal{D}(\Psi(t)) \leq \frac{\mathcal{D}(\Psi^0)}{\mathcal{D}(\Psi^0) + (1 - 2\mathcal{D}(\Psi^0))e^{\kappa t}}.$$

Note that the condition $\mathcal{D}(\Psi^0) < \frac{1}{2}$ is needed to exclude the finite-time blowup of $\mathcal{D}(\Psi(t))$.

(iii) In Refs. [4, 47], the authors provided *"finite dimensional approach"* based on the two-point correlation function $h_{ij} = \langle \psi_i, \psi_j \rangle$ measuring the degree of aggregation. Then, the correlation function h_{ij} with $a_{ik} \equiv 1$ satisfies

$$\frac{\mathrm{d}}{\mathrm{d}t} h_{ij} = \frac{\kappa}{2N} \sum_{k=1}^{N} (h_{ik} + h_{kj})(1 - h_{ij}), \quad 1 \le i, j \le N.$$

By applying dynamical systems theory, we can obtain the desired result. $\square$

Next, we consider a heterogenous ensemble with distinct one-body potentials. In the aforementioned work [47], the authors considered the case of a two-oscillator system whose external potentials are assumed to be the vertical translation of a common potential:

$$V_j(x) = V(x) + \nu_j, \quad x \in \mathbb{R}^d, \quad \nu_j \in \mathbb{R}, \quad j = 1, 2.$$

Then for the system, they exactly find the critical coupling strength for the bifurcation where the system undergoes a transition from the emergence of periodic motion to the existence of equilibrium. More precisely,

(i) $\kappa < \nu := |\nu_1 - \nu_2|$: $h(t)$ is a periodic function with the period $\dfrac{\pi}{\sqrt{\nu^2 - \kappa^2}}$.

(ii) $\kappa = \nu$: $\lim\limits_{t \to \infty} h(t) = -\mathrm{i}$.

(iii) $\kappa > \nu$: $\lim\limits_{t \to \infty} h(t) = -\mathrm{i}\dfrac{\nu}{\kappa} + \sqrt{1 - \dfrac{\nu^2}{\kappa^2}}$.

The emergent dynamics of the SL model with $N \ge 3$ can be studied similarly as in previous section for some restricted class of initial data and system parameters.

Proposition 4.3: [19, 31] *(Heterogeneous ensemble). The following assertions hold.*

(1) Suppose system parameters and initial data satisfy

$$\kappa > 54\mathcal{D}(V) > 0, \quad \mathcal{D}(\Psi^0) < \alpha_2,$$

where α_2 is a larger positive root of $f(x) := 2x^3 - x^2 + \frac{2\mathcal{D}(V)}{\kappa} = 0$, and let $\{\psi_j\}$ be a solution to (2.2). Then, practical aggregation emerges:

$$\lim_{\kappa \to \infty} \limsup_{t \to \infty} \mathcal{D}(\Psi(t)) = 0.$$

(2) *Suppose system parameters and initial data satisfy*

$$\kappa > 4\mathcal{D}(\nu) =: 4 \max_{1 \leq i,j \leq N} |\nu_i - \nu_j|, \quad \mathcal{D}(\Psi^0)^2 < \frac{\kappa + \sqrt{\kappa^2 - 4\kappa\mathcal{D}(\nu)}}{\kappa},$$

(4.2)

and let $\{\psi_j\}$ be a global solution to (2.2). Then for each i, j, there exist a complex number $\alpha_{ij}^\infty \in \mathbb{C}$ such that

$$\lim_{t \to \infty} \langle \psi_i, \psi_j \rangle(t) = \alpha_{ij}^\infty, \quad |\alpha_{ij}^\infty| \leq 1.$$

In other words, state-locking occurs. Moreover, the convergence rate is exponential.

Proof:

(i) Consider the equation:

$$f(x) := 2x^3 - x^2 + \frac{2\mathcal{D}(V)}{\kappa} = 0, \quad x \in [0, \infty), \quad \kappa > 54\mathcal{D}(V).$$

Then, the cubic equation $f = 0$ has a positive local maximum $\frac{2\mathcal{D}(V)}{K}$ and a negative local minimum $\frac{2\mathcal{D}(V)}{K} - \frac{1}{27}$ at $x = 0$ and $\frac{1}{3}$, respectively. Moreover, it has two positive real roots $\alpha_1 < \alpha_2$:

$$0 < \alpha_1 < \frac{1}{3} < \alpha_2 < \frac{1}{2}.$$

Clearly, the roots depend continuously on κ and $\mathcal{D}(V)$, and

$$\lim_{\kappa \to \infty} \alpha_1 = 0, \quad \lim_{\kappa \to \infty} \alpha_2 = \frac{1}{2}.$$

The flow issued from the initial data satisfying $\mathcal{D}(\Psi^0) < \alpha_2$ tends to the set $\{\Psi : \mathcal{D}(\Psi) < \alpha_1\}$ in finite-time. Moreover, this set is positively invariant, i.e.,

$$\mathcal{D}(\Psi(t)) < \alpha_1, \quad t \gg 1.$$

On the other hand one has $\lim_{\kappa \to \infty} \alpha_1(\kappa) = 0$. This yields the desired estimate. Detailed argument can be found in Ref. [19].

(ii) In order to show that the limit of $h_{ij} = \langle \psi_i, \psi_j \rangle$ exists, we consider any two solutions to (2.2) denoted by $\{\psi_j\}$ and $\{\tilde{\psi}_j\}$, and we write

$$h_{ij} = \langle \psi_i, \psi_j \rangle, \quad \tilde{h}_{ij} := \langle \tilde{\psi}_i, \tilde{\psi}_j \rangle,$$

and define the diameter measuring the dissimilarity of two correlation functions:

$$d(\mathcal{H}, \tilde{\mathcal{H}})(t) := \max_{1 \leq i,j \leq N} |h_{ij}(t) - \tilde{h}_{ij}(t)|.$$

Then, we find a differential inequality for $d(\mathcal{H}, \tilde{\mathcal{H}})$:

$$\frac{\mathrm{d}}{\mathrm{d}t} d(\mathcal{H}, \tilde{\mathcal{H}}) \leq -\kappa(1 - \mathcal{D}(\Psi)^2) d(\mathcal{H}, \tilde{\mathcal{H}}), \quad t > 0.$$

Under the assumption (4.2), we show that $d(\mathcal{H}, \tilde{\mathcal{H}})$ tends to zero. Once we establish the zero convergence of $d(\mathcal{H}, \tilde{\mathcal{H}})$, since our system is autonomous, we can choose $\tilde{h}_{ij}$ as $\tilde{h}_{ij}(t) = h_{ij}(t + T)$ for any $T > 0$. By discretizing the time $t \in \mathbb{R}_+$ as $n \in \mathbb{Z}_+$, one can deduce that $\{h_{ij}(n)\}_{n \in \mathbb{Z}_+}$ becomes a Cauchy sequence in the complete space $\{z \in \mathbb{C} : |z| \leq 1\}$. Thus, we find the desired complex number α_{ij}^∞. $\qquad\square$

4.1.2. Network structure

For the interplay between emergent behaviors and network structures, the authors in Ref. [48] considered the following three types of network structures: cooperative, competitive and cooperative-competitive networks depending on their signs:

(i) Cooperative: $a_{ik} > 0$ for all $i, k = 1, \cdots, N$.

(ii) Competitive: $a_{ik} < 0$ for all $i, k = 1, \cdots, N$.

(iii) Cooperative-competitive: $a_{ik} = (-1)^{i+k}$ for $i < k$ and $a_{ki} = -a_{ik}$.

For the cooperative network a_{ik}, all values of a_{ik} have positive values so that we can expect complete state aggregation. Here, we associate statistical quantities for the cooperative network $\{a_{ik}\}$:

$$d(\mathcal{A}) := \max_{1 \leq i,j,k \leq N} |a_{ik} - a_{jk}|, \quad a_m^c := \min_{1 \leq i \leq N} \frac{1}{N} \sum_{k=1}^{N} a_{ik}, \quad a_M := \max_{1 \leq i,j \leq N} a_{ij}.$$

On the other hand, for the competitive network, all values of a_{ik} are assumed to be negative and they considered the simplest case $a_{ik} \equiv -1$. In this case, each oscillator would exhibit repulsive behaviors. Lastly for the cooperative-competitive network, some of a_{ik} are positive and some are negative. Hence, we expect interesting dynamical patterns other than aggregation, such as periodic orbit or bi-polar aggregation. The arguments above are summarized in the following theorem.

Theorem 4.4: [48] *Let $\{\psi_j\}$ be a global solution to (2.2).*

(1) Suppose that a_{ik} and initial data satisfy

$$a_{ik} > 0, \quad d(\mathcal{A}) < a_m^c, \quad \mathcal{D}(\Psi^0)^2 < \frac{2(a_m^c - d(\mathcal{A}))}{a_M}.$$

Then, system (2.2) exhibits complete state aggregation.

(2) Suppose that a_{ik} and initial data satisfy

$$a_{ik} \equiv -1, \quad \rho^0 > 0.$$

Then, a solution to (2.2) tends to the splay state.

(3) Suppose for $N = 4$ that a_{ik} and initial data satisfy

$$a_{ik} = (-1)^{i+k}, \quad 2 + h_{12}^0 + h_{14}^0 + h_{23}^0 + h_{34}^0 < h_{13}^0 + h_{24}^0.$$

Then, we have

$$\lim_{t \to \infty} h_{ij}(t) = (-1)^{i+j}.$$

4.2. *Standing wave solution*

In this subsection, we consider a specific type of a solution whose shape is invariant under the flow, namely, a standing wave solution. We begin with the ansatz for ψ_i:

$$\psi_j(x, t) = u(x)e^{-iEt} \quad \text{with} \quad \|u\| = 1, \quad j = 1, \cdots, N, \tag{4.3}$$

where E is a real number. We substitute the ansatz (4.3) into (2.2) with an identical harmonic potential $V_i(x) = |x|^2$ to derive

$$-\Delta u + |x|^2 u = Eu, \quad x \in \mathbb{R}^d. \tag{4.4}$$

In what follows, we consider the one-dimensional $d = 1$ case and generalization to the multi-dimensional case can be constructed from the tensor product of a one-dimensional solution. For the one-dimensional case, Eq. (4.4) becomes

$$-u_{xx} + x^2 u = Eu, \quad x \in \mathbb{R}. \tag{4.5}$$

Then, it is well-known that the Eq. (4.5) has eigenvalues E_k and orthonormal eigenfunctions (or constant multiple of the Hermite functions) u_k:

$$E_k = 2k + 1, \quad u_k(x) = \left(\frac{1}{\sqrt{\pi} 2^k k!} \right)^{1/2} e^{-x^2/2} H_k(x), \quad k = 0, 1, \cdots,$$
$$\tag{4.6}$$

where H_k is the k-th Hermite polynomial defined by

$$H_k(x) = (-1)^k e^{x^2} \frac{d^k}{dx^k} e^{-x^2}, \quad k = 0, 1, \cdots.$$

We now consider the following Cauchy problem:

$$
\begin{cases}
iu_t = -u_{xx} + x^2 u, & x \in \mathbb{R}, \ t > 0, \\
u(x,0) = u_0(x) = \displaystyle\sum_{k=0}^{\infty} a_k u_k(x),
\end{cases}
\tag{4.7}
$$

where $\{u_k\}$ is defined in (4.6) and $\{a_n\}$ is a ℓ^2-sequence of complex numbers. Then, it can be easily seen that the solution $u(x,t)$ of the Cauchy problem (4.7) is given as follows:

$$
u(x,t) = \sum_{k=0}^{\infty} a_k u_k(x) e^{-i(2k+1)t}.
$$

Next, we study the stability issue for the two types of the standing wave solutions whose existence is guaranteed by the previous argument:

(I) $\psi_j(x,t) = u_k(x) e^{-i(2k+1)t}$ for $j = 1, 2, \cdots, N$.

(II) $\psi_1(x,t) = -u_k(x) e^{-i(2k+1)t}$ and $\psi_j = u_k(x) e^{-i(2k+1)t}$ for $j \neq 1$,

where u_k is given by the formula (4.6).

Note that the family (I) corresponds to the completely aggregated state, whereas the family (II) corresponds to bi-polar state. As in Theorem 4.2, bi-polar state is unstable. Below, under the initial condition for which complete state aggregation occurs, the SL model becomes stable.

Theorem 4.5: [49] *The family (I) is stable in the following sense: for all $\varepsilon > 0$, there exists $\delta > 0$ such that*

$$
\|\psi_j^0 - u_k\| < \delta \quad \Longrightarrow \quad \lim_{t \to \infty} \|\psi_j(t) - u_k e^{-i(2k+1)t}\| < \varepsilon.
$$

Remark 4.6: Note that we cannot expect an asymptotic stability of $u_k(x) e^{-i(2k+1)t}$. In fact, we consider the following form of solution:

$$
\psi_j(x,t) = (1-a) u_k(x) e^{-i(2k+1)t} + b u_{k+1}(x) e^{-i(2k+3)t} \quad \text{for } j = 1, 2, \cdots, N,
$$

where $|1-a|^2 + |b|^2 = 1$. Then, it is easy to check that the above ψ_j is a solution to (2.2) and

$$
\|\psi_j(x,t) - u_k(x) e^{-i(2k+1)t}\|^2 = |a|^2 + |b|^2 = 2\mathrm{Re}(a).
$$

4.3. Numeric scheme

In Ref. [11], the authors consider nonlinearly coupled Schrödinger-Lohe type system by employing cubic nonlinearity so that the model would be reduced to the Gross-Pitaevskii equation when the coupling is turned off:

$$
\begin{cases}
\mathrm{i}\partial_t \psi_j = -\dfrac{1}{2}\Delta\psi_j + V_j\psi_j + \displaystyle\sum_{k=1}^{N}\beta_{jk}|\psi_k|^2\psi_j \\[2mm]
\qquad\quad + \dfrac{\mathrm{i}\kappa}{2N}\displaystyle\sum_{k=1}^{N} a_{jk}\left(\psi_k - \dfrac{\langle\psi_j,\psi_k\rangle}{\langle\psi_j,\psi_j\rangle}\psi_j\right), \\[2mm]
\psi_j(x,0) = \psi_j^0(x), \quad (x,t)\in\mathbb{R}^d\times\mathbb{R}_+, \quad j = 1,\cdots,N.
\end{cases}
\tag{4.8}
$$

Next, we discuss an efficient and accurate numerical method for discretizing (4.8). Several numerical examples will be carried out and compared with corresponding analytical results shown in previous section. Due to the external trapping potential $V_j(x)$ ($j = 1,\cdots,N$), the wave functions ψ_j ($j = 1,\cdots,N$) decay exponentially fast as $|x|\to\infty$. Therefore, it suffices to truncate the problem (4.8) into a sufficiently large bounded domain $\mathcal{D}\subset\mathbb{R}^d$ with periodic boundary condition (BC). The bounded domain $\mathcal{D}$ is chosen as a box $[a,b]\times[c,d]\times[e,f]$ in 3D, a rectangle $[a,b]\times[c,d]$ in 2D, and an interval $[a,b]$ in 1D.

4.3.1. A time splitting Crank-Nicolson spectral method

First, we begin with the description of (4.8) combining a time splitting spectral method and the Crank-Nicolson method. Choose $\Delta t > 0$ as the time step size and denote time steps $t_n := n\Delta t$ for $n \geq 0$. From time $t = t_n$ to $t = t_{n+1}$, the GPL is solved in three splitting steps. One solves first

$$
\mathrm{i}\partial_t\psi_j = -\frac{1}{2}\Delta\psi_j, \quad x\in\mathcal{D}, \quad j = 1,\cdots,N,
\tag{4.9}
$$

with periodic BC on the boundary $\partial\mathcal{D}$ for the time step of length Δt, then solves

$$
\mathrm{i}\partial_t\psi_j = V_j\psi_j + \sum_{k=1}^{N}\beta_{jk}|\psi_k|^2\psi_j, \quad j = 1,\cdots,N,
\tag{4.10}
$$

for the same time step, and finally solves

$$
\mathrm{i}\partial_t\psi_j = \frac{\mathrm{i}\kappa}{2N}\sum_{k=1}^{N} a_{jk}\left(\psi_k - \frac{\langle\psi_j,\psi_k\rangle}{\langle\psi_j,\psi_j\rangle}\psi_j\right), \quad j = 1,\cdots,N,
\tag{4.11}
$$

for the same time-step. The linear subproblem (4.9) is discretized in space
by the Fourier pseudospectral method and integrated in time analytically in
the phase space [5–7,10]. For the nonlinear subproblem (4.10), it conserves
$|\psi_k|^2$ pointwise in time, i.e. $|\psi_k(x,t)|^2 \equiv |\psi_k(x,t_n)|^2$ for $t_n \leq t \leq t_{n+1}$ and
$k = 1, \ldots, N$ [5–7,10]. Thus it collapses to a linear subproblem and can
be integrated in time analytically [5–7,10]. For the nonlinear subproblem
(4.11), due to the presence of the Lohe term involving κ, it cannot be
integrated analytically (or explicitly) in the way for the standard GPE [5,7].
Therefore, we will apply a Crank-Nicolson scheme [8] to further discretize
the temporal derivate of (4.11).

To simplify the presentation, we will only provide the scheme for 1D.
Generalization to $d > 1$ is straightforward for tensor grids. To this end, we
choose the spatial mesh size as $\Delta x = \frac{b-a}{M}$ with M a even positive integer,
and let the grid points be

$$x_\ell = a + \ell \Delta x, \qquad \ell = 0, \cdots, M.$$

For $1 \leq j \leq N$ denote $\psi_{j,\ell}^n$ as the approximation of $\psi_j(x_\ell, t_n)$ $(0 \leq \ell \leq M)$ and $\boldsymbol{\psi}_j^n$ as the solution vector with component $\psi_{j,\ell}^n$. Combining the
time splitting (4.9)–(4.11) via the Strang splitting and the Crank-Nicolson
scheme for (4.11), a second order *Time Splitting Crank-Nicolson Fourier
Pseudospectral* (TSCN-FP) method to solve GPL on $\mathcal{D}$ reads as:

$$\psi_{j,\ell}^{(1)} = \sum_{p=-M/2}^{M/2-1} e^{-i\Delta t\, \mu_p^2/4} \widehat{(\boldsymbol{\psi}_j^n)}_p\, e^{i\mu_p(x_\ell-a)},$$

$$\psi_{j,\ell}^{(2)} = e^{-i\Delta t\left(V_j(x_\ell) + \sum_{k=1}^N \beta_{jk}|\psi_{k,\ell}^{(1)}|^2\right)/2}\, \psi_{j,\ell}^{(1)},$$

$$i\frac{\psi_{j,\ell}^{(3)} - \psi_{j,\ell}^{(2)}}{\Delta t} = \frac{i\kappa}{2N}\sum_{k=1}^N a_{jk}\left[\psi_{k,\ell}^{(\frac{5}{2})} - \frac{\langle \boldsymbol{\psi}_j^{(\frac{5}{2})}, \boldsymbol{\psi}_k^{(\frac{5}{2})}\rangle_{\Delta x}}{\langle \boldsymbol{\psi}_j^{(\frac{5}{2})}, \boldsymbol{\psi}_j^{(\frac{5}{2})}\rangle_{\Delta x}}\, \psi_{j,\ell}^{(\frac{5}{2})}\right], \qquad (4.12)$$

$$\psi_{j,\ell}^{(4)} = e^{-i\Delta t\left(V_j(x_\ell) + \sum_{k=1}^N \beta_{jk}|\psi_{k,\ell}^{(3)}|^2\right)/2}\, \psi_{j,\ell}^{(3)}, \quad 0 \leq \ell \leq M,$$

$$\psi_{j,\ell}^{n+1} = \sum_{p=-M/2}^{M/2-1} e^{-i\Delta t\, \mu_p^2/4} \widehat{(\boldsymbol{\psi}_j^{(4)})}_p\, e^{i\mu_p(x_\ell-a)}, \quad j = 1, \cdots, N.$$

Here, $\mu_p = \frac{p\pi}{b-a}$, $\widehat{(\boldsymbol{\psi}_j^n)}_p$ and $\widehat{(\boldsymbol{\psi}_j^{(4)})}_p$ $(p = -\frac{M}{2}, \cdots, \frac{M}{2})$ are the discrete
Fourier transform coefficients of the vectors $\boldsymbol{\psi}_j^n$ and $\boldsymbol{\psi}_j^{(4)}$ $(j = 1, \cdots, N)$,

respectively. Moreover,

$$\psi_{j,\ell}^{(\frac{5}{2})} =: \frac{1}{2}\left(\psi_{j,\ell}^{(3)} + \psi_{j,\ell}^{(2)}\right), \qquad \langle \boldsymbol{\psi}_j^{(\frac{5}{2})}, \boldsymbol{\psi}_k^{(\frac{5}{2})} \rangle_{\Delta x} =: \Delta x \sum_{\ell=0}^{M-1} \psi_{j,\ell}^{(\frac{5}{2})} \, \bar{\psi}_{k,\ell}^{(\frac{5}{2})}.$$

Although the Crank-Nicolson step (4.12) is fully implicit, it can be either solved efficiently by Krylov subspace iteration method with proper pre-conditioner [3] or the fixed-point iteration method with a stabilization parameter [9]. In addition, TSCN-FP is of spectral accuracy in space and second-order accuracy in time. By following the standard procedure, it is straightforward to show that the TSCN-FP conserve mass of each component in discrete level, i.e., $\|\boldsymbol{\psi}_j^n\|_{l^2}^2 := \langle \boldsymbol{\psi}_j^n, \boldsymbol{\psi}_j^n \rangle_{\Delta x} \equiv \|\boldsymbol{\psi}_j^0\|_{l^2}^2$ for $n \geq 0$ and $j = 1, 2, \ldots, N$. We omit the details here for brevity.

4.3.2. *Numerical results*

In this subsection, we apply the TSCN-FP schemes proposed in the previous subsection to simulate some interesting dynamics. For our simulation, we choose

$$\beta = 1, \quad \Delta t = 2 \times 10^{-4}, \quad \mathcal{D} = [-12, 12]^d, \quad d = 1, 2.$$

The potentials and initial data are chosen as follows:

$$V_j(x) = \pi^2 \alpha_j^2 \, |x|^2, \quad \psi_j^0 = \sqrt{a_j/\pi} \, e^{-a_j|x - x_0^j|^2}.$$

Here, α_j and x_0^j are real constants to be given later. In fact, complete state aggregation and practical aggregation estimates do not depend on the form of the initial data and the relative L^2-distances of the initial data play a crucial role. However, when we deal with the center-of-mass x_c, we used the Gaussian initial data so that they have the symmetric form (see Remark 4.2 in Ref. [11]). For the numerical experiment, we introduce the following quantities (see Ref. [11] for details):

$$R(t) := \mathrm{Re}\langle \psi_1, \psi_2 \rangle(t), \quad \mathcal{R}_{ijk\ell} := \frac{(1 - h_{ij})(1 - h_{k\ell})}{(1 - h_{i\ell})(1 - h_{kj})},$$

$$\mathcal{B} := (\beta_{ij})_{1 \leq i,j \leq N}, \quad x_c^j(t) := \int_{\mathbb{R}} x |\psi_j(x, t)|^2 dx,$$

$$\mathcal{E}[\Psi] := \sum_{j=1}^{N} \int_{\mathbb{R}^d} \left[\frac{1}{2}|\nabla \psi_j|^2 + V_j|\psi_j|^2 + \frac{1}{2}\sum_{k=1}^{N} \beta_{jk}|\psi_k|^2|\psi_j|^2 \right] dx.$$

Example 4.7: *Here, we consider the two-component system in 1D, i.e., we take $N = 2$ and $d = 1$ in (2.2). To this end, we take $(x_1^0, x_2^0) = (2.5, -5)$ and consider the following two cases: for $j = 1, 2$,*

> ***Case 1.*** *fix $\alpha_j = \beta_{j\ell} = 1$ ($\ell = 1, 2$) and vary $\kappa = 0, 2, 20$.*
> ***Case 2.*** *fix $\alpha_j = j$, $\beta_{12} = \beta_{21} = 1$, $\beta_{11} = 4\beta_{22} = 2$ and vary $\kappa = 0, 2, 10, 20$.*

*Figures 1 and 2 depict the time evolution of the quantity $1 - R(t)$ (where $R(t)$ is the real part of the correlation function $h_{12}(t)$), the center of mass $x_c^j(t)$, the component mass $\|\psi_j\|^2$ and the total energy $\mathcal{E}(t)$ for **Case 1** and **Case 2**, respectively. From these figures and other numerical experiments not shown here for brevity, we can see the following observations:*

(i). For all cases, we observe that the mass is conserved along time.

(ii). If the Lohe coupling is off, i.e., $\kappa = 0$, both the mass and energy are conserved well, and the center of mass $(x_c^1(t), x_c^2(t))$ are periodic in time with the same period. In addition, for the identical case, i.e., $\mathcal{B} = J_2$ and $V_1(x) = V_2(x)$, $R(t)$ is conserved for identical case.

(iii). If the Lohe coupling is on, i.e., $\kappa > 0$, the phenomena become complicated. The energy is no longer conserved, indeed it decays to some value for large κ while it oscillates for small κ.

(iv). Moreover, for the identical case, $R(t)$ converges exponentially to 1, which coincides with the theoretical results. Thus, complete state aggregation occurs in this case. After complete state aggregation, $\|\psi_1(x, t) - \psi_2(x, t)\|_\infty$ will converge to zero and the center of mass $x_c^1(t)$ and $x_c^2(t)$ will become the same and swing periodically along the line connecting $-\bar{x}_c^0$ and $\bar{x}_c^0$ (here, $\bar{x}_c^0 := (x_c^1(0) + x_c^2(0))/2$).

(v). Furthermore, for the non-identical case, i.e., $\mathcal{B} \neq J_2$ and $V_1(x) \neq V_2(x)$, $R(t)$ does not converge to 1, i.e., complete state aggregation cannot occur. However, for large κ, $R(t)$ indeed converges to some definite constant $R_\infty < 1$. The larger κ, the smaller value $1 - R_\infty$. Meanwhile, $|x_c^1(t) - x_c^2(t)|$ also converges to zero, which could be also justified in a similar process as shown in Corollary 4.1 of Ref. [11].

Example 4.8: *Here, we consider the six-component system in 2D, i.e., we take $N = 6$ and $d = 2$ in (2.2). To this end, we only consider the identical case, i.e., we choose $\alpha_j = 1 = \beta_{j\ell} = 1$ ($j, \ell = 1, \cdots, 6$). Let $\kappa = 20$, we consider four cases of initial setups:*

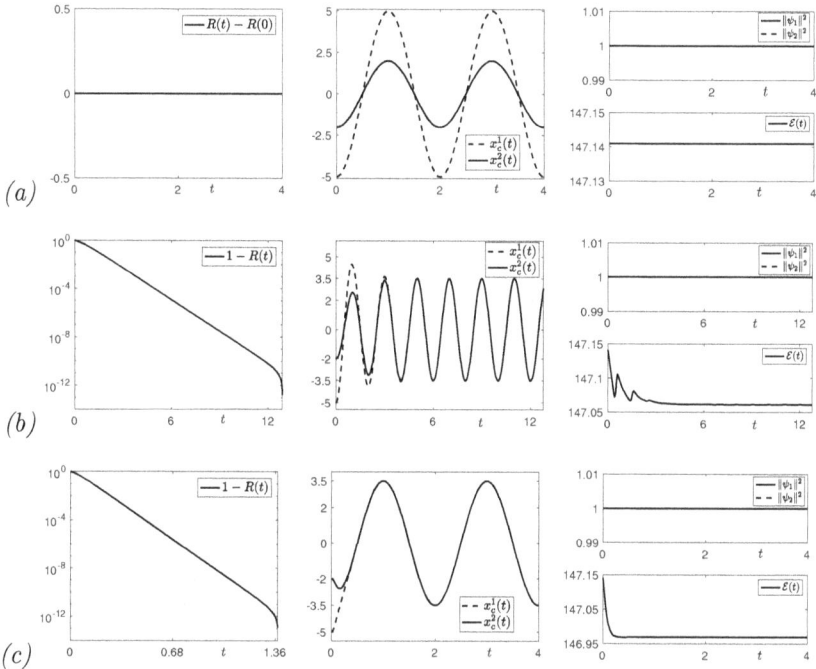

Fig. 1.   Time evolution of the quantity $1 - R(t)$ (left), the center of mass $x_c^j(t)$ (middle), and the component mass $\|\psi_j\|^2$ and the total energy $\mathcal{E}(t)$ (right) for **Case 1** in Example 4.7 for $\kappa = 0, 2, 20$ (top to bottom).

Case 3. $x_0^j = \big(6\cos((j-1)\pi/3), 6\sin((j-1)\pi/3)\big), \quad j = 1, \cdots, 6.$
Case 4. $x_0^j = \big(2 + 4\cos(j\pi/3 - \pi/12), 2 + 4\sin(j\pi/3 - \pi/12)\big),$
$j = 1, \cdots, 6.$
Case 5. $x_0^j = \big(6\cos((j-1)\pi/5), 6\sin((j-1)\pi/5)\big), \quad j = 1, \cdots, 6.$
Case 6. *Random location:*

$$x_0^1 = (3.4707, 2.7526), \quad x_0^2 = (-0.8931, 1.9951),$$
$$x_0^3 = (0.1809, -1.1538), \quad x_0^4 = (0.0937, -5.8995),$$
$$x_0^5 = (-2.9235, -2.4171), \quad x_0^6 = (-3.6423, 4.3714).$$

*For **Case 3**–**Case 6**, Fig. 3 illustrates the trajectory and time evolution of the center of mass $x_c^j(t) =: (x_{c1}^j(t), x_{c2}^j(t))$, Fig. 4 depicts the time evolution of $|\mathcal{R}_{1256}(t) - \mathcal{R}_{1256}(0)|$, $|\mathcal{R}_{2456}(t) - \mathcal{R}_{2456}(0)|$ and $|\mathcal{R}_{3456}(t) - \mathcal{R}_{3456}(0)|$, and Fig. 5 shows the contour plots of $|\psi_1(x, t)|^2$ at different times. From these figures and other numerical experiments not shown here for brevity, we can see the following observations.*

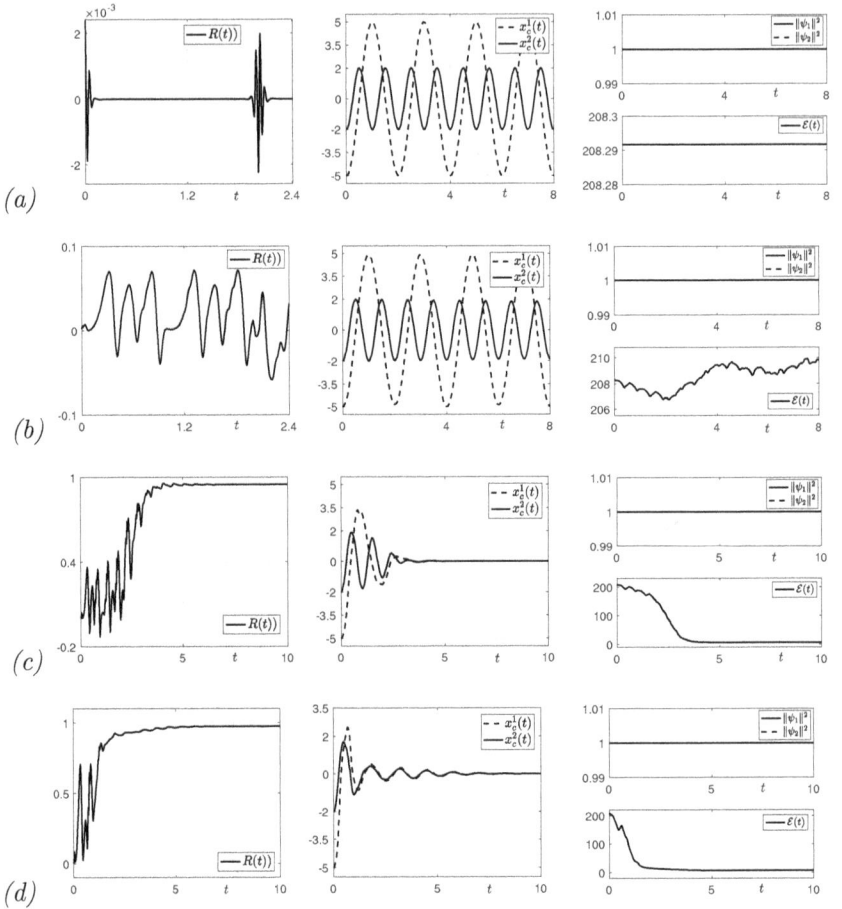

Fig. 2.   Time evolution of the quantity $1 - R(t)$ (left), the center of mass $x_c^j(t)$ (middle), and the component mass $\|\psi_j\|^2$ and the total energy $\mathcal{E}(t)$ (right) for **Case 2** in Example 4.7 for $\kappa = 0, 2, 10, 20$ (top to bottom).

(i). Complete state aggregation occurs for all cases.

(ii). All the center of mass $x_c^j(t)$ $(j = 1, \cdots, 6)$ will converge to the same periodic function $\bar{x}_c(t)$, which swings exactly along the line connecting the points $(-\bar{x}_{c1}^0, -\bar{x}_{c2}^0)$ and $(\bar{x}_{c1}^0, \bar{x}_{c2}^0)$ which are defined as the average of the initial center of mass of the six oscillators:

$$(\bar{x}_{c1}^0, \bar{x}_{c2}^0) := \left(\frac{1}{6} \sum_{j=1}^{6} x_{c1}^j(0), \frac{1}{6} \sum_{j=1}^{6} x_{c2}^j(0) \right).$$

Thus, when $\bar{x}^0_{c1} = \bar{x}^0_{c2} = 0$, the center of mass will stay steady at the origin (cf. Fig. 3(a)), which also agrees with the conclusion in Remark 4.2 of Ref. [11].

(iii). Before complete state aggregation, all density profiles $|\psi_j(x,t)|^2$ ($j = 1, \cdots, 6$) will evolve similarly, i.e., the same dynamical pattern as those shown in Fig. 5 for $|\psi_1|^2$ (only differ from the 'color', i.e., the more blurred humps imply the centers of the other five components, while the lighter one shows the one of the current component). While after complete state aggregation (around $t = 0.4$, which corresponds to the moment the center of mass x_c^j ($j = 1, \cdots, 6$) meet together in Fig. 3), all $\psi_j(x,t)$ (hence also for all density profiles) will converge to the same function, whose density changes periodically in time (as shown in columns 4–6 in Fig. 5, which also indicate the periodic dynamics for the center of mass that illustrated in Fig. 3). In addition, before complete state aggregation, although numerical schemes cannot conserve the cross-ratio like quantities $\mathcal{R}_{ijkl}(t)(1 \leq i,j,k,l \leq 6)$ in discretized level, the difference of those quantities from their initial ones are still small (cf. Fig. 4).

5. Conclusion

In this chapter, we have reviewed state-of-the-art results on the collective behaviors for two Lohe type aggregation models (the Lohe tensor model and the Schrödinger-Lohe model). The former deals with the aggregation dynamics of tensors with the same rank and size, and it turns out to be a generalized model for previously known first-order aggregation models such as the Kuramoto model, the swarm sphere model and the Lohe matrix model. Of course, the Lohe tensor model can be reduced to aggregation models on the hermitian sphere and non-square matrix group which are not known in previous literature. For the collective dynamics, we adopt two concepts of aggregation (complete state aggregation and practical aggregation). When all state aggregates to the same state, we call it as complete state aggregation. This phenomenon occurs for a homogeneous ensemble in which all particles follow the same free flow. In contrast, when states are governed by different free flows, complete state aggregation cannot occur. Of course, the rate of state change (we call it velocity) can aggregate to the same value. At present, this strong estimate is not available for the aforementioned models yet.

The latter describes the collective aggregation of the coupled

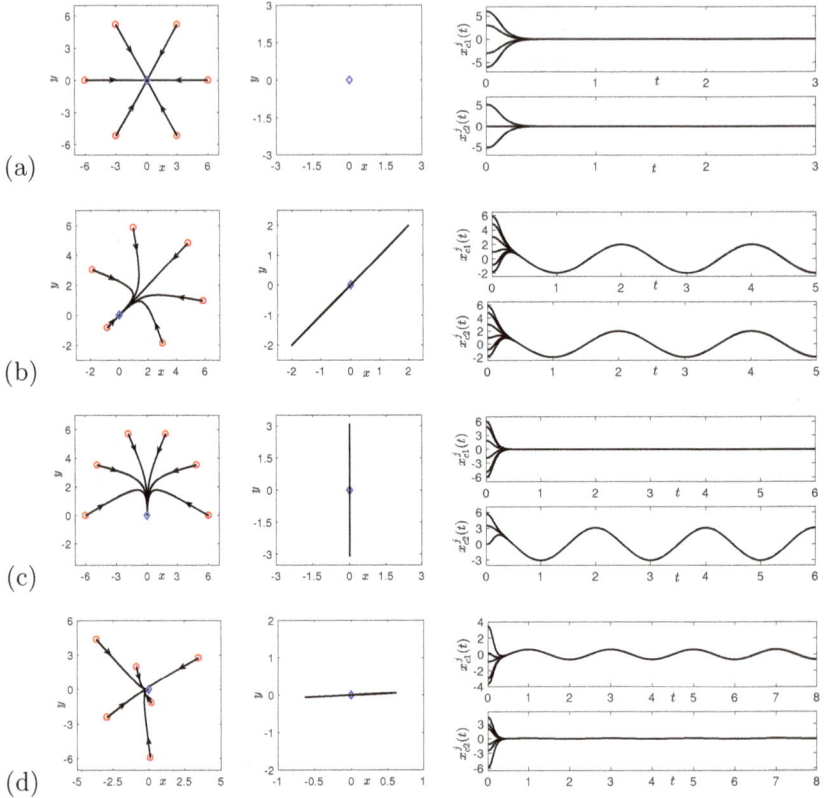

(a)

(b)

(c)

(d)

Fig. 3.   First two columns: trajectory of center of mass $x_c^j(t)$ in $t \in [0, t_c]$ and $t \in [t_c, 10]$ ($t_c = 1.5$ for first row while 0.5 for the others). The third column: time evolution of $x_{c1}^j(t)$ and $x_{c2}^j(t)$) (right). $\circ$ denotes location of $x_c^j(0)$, while $\diamond$ denotes the one of $x_c^j(t_c)$.

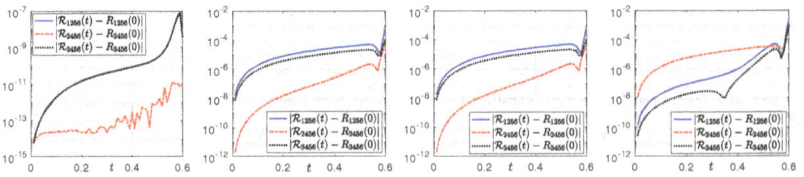

Fig. 4.   Time evolution of $|\mathcal{R}_{1256}(t) - \mathcal{R}_{1256}(0)|$, $|\mathcal{R}_{2456}(t) - \mathcal{R}_{2456}(0)|$ and $|\mathcal{R}_{3456}(t) - \mathcal{R}_{3456}(0)|$ for **Cases 3–6** (left to right).

Schrödinger equations for wave functions. For a special case, it can be reduced to the swarm sphere and Kuramoto models. For this model, we can also adopt the same universal approaches (Lyapunov functional approach and dynamical systems theory approach for two-point correlation

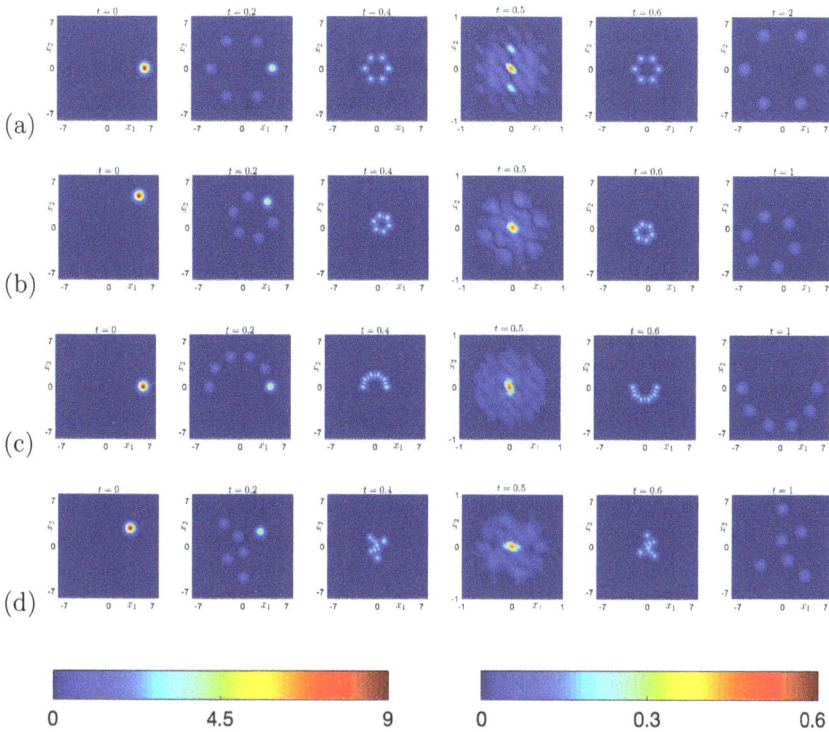

Fig. 5. Contour plots of $|\psi_1(x,t)|^2$ at different time t for **Cases 3–5** in Example 4.8 (the top 4 rows) and color bars of the contour plots at $t = 0.5$ (bottom left) and other time t (bottom right).

functions). Similar to the Lohe tensor model, this model exhibits the same aggregation phenomena for homogeneous and heterogeneous ensembles. As we have already mentioned, we do not have a complete theory for dealing with a heterogeneous ensemble except for a weak aggregation estimate (practical aggregation). Thus, we would say that our reviewed results for a heterogeneous ensemble are still far from complete. This will be an interesting research direction for those who are interested in collective dynamics.

Acknowledgments

The content of this chapter is based on a lecture given by the first author at the Institute of Mathematical Sciences of National University of Singapore in December 2019. The authors acknowledge the Institute of Mathematical Sciences for generous support and especially thank Prof. Weizhu Bao for the invitation and his warm hospitality. The work of S.-Y. Ha is supported by the NRF grant (2020R1A2C3A01003881) and the work of D. Kim was supported by the National Research Foundation of Korea (NRF) grant funded by the Korea government (MSIT) (No. 2021R1F1A1055929).

References

1. J. A. Acebron, L. L. Bonilla, C. J. P. Pérez Vicente, F. Ritort and R. Spigler, "The Kuramoto model: a simple paradigm for synchronization phenomena", Rev. Mod. Phys., **77** (2005), 137–185.
2. G. Albi, N. Bellomo, L. Fermo, S.-Y. Ha, J. Kim, L. Pareschi, D. Poyato and J. Soler, "Vehicular traffic, crowds and swarms: From kinetic theory and multiscale methods to applications and research perspectives", Math. Models Methods Appl. Sci., **29** (2019), 1901–2005.
3. X. Antoine and R. Duboscq, "Robust and efficient preconditioned Krylov spectral solvers for computing the ground states of fast rotating and strongly interacting Bose-Einstein condensates", J. Comput. Phys., **258** (2014), 509–523.
4. P. Antonelli and P. Marcati, "A model of synchronization over quantum networks", J. Phys. A., **50** (2017), 315101.
5. W. Bao, "Ground states and dynamics of multicomponent Bose-Einstein condensates", Multiscale Model. SImul., **2** (2004), 210–236.
6. W. Bao and Y. Cai, "Mathematical models and numerical methods for spinor Bose-Einstein condensates", Commun. Comput. Phys., **24** (2018), 899–965.
7. W. Bao and Y. Cai, "Mathematical theory and numerical methods for Bose-Einstein condensation", Kinet. Relat. Models, **6** (2013), 1–135.
8. W. Bao and Y. Cai, "Uniform error estimates of finite difference methods for the nonlinear Schrödinger equation with wave operator", SIAM J. Numer. Anal., **50** (2012), 492–521.
9. W. Bao, I. Chern and F. Lim, "Efficient and spectrally accurate numerical methods for computing ground and first excited states in Bose-Einstein condensates", J. Comput. Phys., **219** (2006), 836–854.
10. W. Bao, D. Jaksch and P. A. Markowich, "Numerical solution of the Gross-Pitaevskii equation for Bose-Einstein condensation", J. Comput. Phys., **187** (2003), 318–342.
11. W. Bao, S.-Y. Ha, D. Kim and Q. Tang, "Collective synchronization of the multi-component Gross-Pitaevskii-Lohe system", Phys. D, **400** (2019), 132158.
12. J. Bronski, T. Carty and S. Simpson, "A matrix valued Kuramoto model", J. Stat. Phys., **178** (2020), 595–624.
13. J. Buck and E. Buck, "Biology of synchronous flashing of fireflies", Nature, **211** (1966), 562–564.
14. J. Byeon, S.-Y. Ha, G. Hwang and H. Park, "Emergent behaviors of the kinetic Lohe Hermitian sphere model", Archived as arXiv:2104.13036.
15. J. Byeon, S.-Y. Ha and H. Park, "Asymptotic interplay of states and

adapted coupling gains in the Lohe hermitian sphere model", Archived as arXiv:2101.03450.

16. J. A. Carrillo, Y.-P. Choi, C. Totzeck and O. Tse, "An analytical framework for consensus-based global optimization method", Math. Models Methods Appl. Sci., **28** (2018), 1037–1066.

17. J. A. Carrillo, S. Jin, L. Li and Y. Zhu, "A consensus-based global optimization method for high dimensional machine learning problems", ESAIM: COCV, **27** (2021), S5.

18. D. Chi, S.-H. Choi and S.-Y. Ha, "Emergent behaviors of a holonomic particle system on a sphere", J. Math. Phys., **55** (2014), 052703.

19. J. Cho, S.-H. Choi and S.-Y. Ha, "Practical quantum synchronization for the Schrödinger-Lohe system", J. Phys. A, **49** (2016), 205203.

20. S.-H. Choi and S.-Y. Ha, "Time-delayed interactions and synchronization of identical Lohe oscillators", Quart. Appl. Math., **74** (2016), 297–319.

21. S.-H. Choi and S.-Y. Ha, "Large-time dynamics of the asymptotic Lohe model with a small time-delay", J. Phys. A., **48** (2015), 425101.

22. S.-H. Choi and S.-Y. Ha, "Quantum synchronization of the Schrödinger-Lohe model", J. Phys. A, **47** (2014), 355104.

23. S.-H. Choi and S.-Y. Ha, "Complete entrainment of Lohe oscillators under attractive and repulsive couplings", SIAM J. Appl. Dyn. Syst., **13** (2014), 1417–1441.

24. F. Cucker and S. Smale, "Emergent behavior in flocks", IEEE Trans. Automat. Control, **52** (2007), 852–862.

25. P. Degond, A. Frouvelle, S. Merino-Aceituno and A. Trescases, "Quaternions in collective dynamics", Multiscale Model. Simul., **16** (2018), 28–77.

26. L. DeVille, "Synchronization and stability for quantum Kuramoto", J. Stat. Phys., **174** (2019), 160–187.

27. F. Dörfler and F. Bullo, "Synchronization in complex networks of phase oscillators: A survey", Automatica, **50** (2014), 1539–1564.

28. M. Fornasier, H. Huang, L. Pareschi and P. Sunnen, "Consensus-based optimization on the sphere II: Convergence to global minimizers and machine learning." Archived as arXiv:2001.11988.

29. Fornasier, M., Huang, H., Pareschi, L. and Sunnen, P.: *Consensus-based optimization on hypersurfaces: Well-posedness and mean-field limit.* Math. Models Methods Appl. Sci. **30** (2020), 2725–2751.

30. F. Golse and S.-Y. Ha, "A mean-field limit of the Lohe matrix model and emergent dynamics", Arch. Rational Mech. Anal., **234** (2019), 1445–1491.

31. S.-Y. Ha, G. Hwang and D. Kim, "Remarks on the stability properties for the continuous and discrete Schrödinger-Lohe system", In preparation.

32. S.-Y. Ha, G. Hwang and H. Park, "Emergent behaviors of Lohe Hermitian sphere particles under time-delay interactions", Netw. Heterog Media, (2021).

33. S.-Y. Ha, M. Kang and H. Park, "Collective behaviors of the Lohe Hermitian sphere model with inertia", Commun. Pure Appl. Anal., (2021).

34. S.-Y. Ha, M. Kang and H. Park, "Emergent dynamics of the Lohe Hermitian sphere model with frustration", J. Math. Phys., **62** (2021), 052701.

35. S.-Y. Ha and D. Kim, "Emergence of synchronous behaviors for the

Schrödinger-Lohe model with frustration", Nonlinearity, **32** (2019), 4609–4637.

36. S.-Y. Ha and D. Kim, "Uniform stability and emergent dynamics of particle and kinetic Lohe matrix models", Under review.

37. S.-Y. Ha, D. Kim, J. Lee and S. E. No, "Particle and kinetic models for swarming particles on a sphere and stability properties", J. Stat. Phys., **174** (2019), 622–655.

38. S.-Y. Ha, D. Kim and H. Park, "Existence and emergent dynamics of quadratically separable states to the Lohe tensor model", Archived as arXiv:2103.17029.

39. S.-Y. Ha, D. Kim and H. Park, "On the completely separable state for the Lohe tensor model", J. Stat. Phys., **183** (2021).

40. S.-Y. Ha, D. Ko, J. Park and X. Zhang, "Collective synchronization of classical and quantum oscillators", EMS Surv. Math. Sci., **3** (2016), 209–267.

41. S.-Y. Ha and H. Park, "A gradient flow formulation of the Lohe matrix model with a high-order polynomial coupling", To appear in J. Stat. Phys.

42. S.-Y. Ha and H. Park, "Emergent behaviors of the generalized Lohe matrix model", Discrete Contin. Dyn. Syst. B, **26** (2021), 4227–4261.

43. S.-Y. Ha and H. Park, "Complete aggregation of the Lohe tensor model with the same free flow", J. Math. Phys., **61** (2020), 102702.

44. S.-Y. Ha and H. Park, " From the Lohe tensor model to the Lohe Hermitian sphere model and emergent dynamics", SIAM J. Appl. Dyn. Syst., **19** (2020), 1312–1342.

45. S.-Y. Ha and H. Park, "Emergent behaviors of Lohe tensor flock", J. Stat. Phys., **178** (2020), 1268–1292.

46. S.-Y. Ha and S. W. Ryoo, "On the emergence and orbital Stability of phase-locked states for the Lohe model", J. Stat. Phys., **163** (2016), 411–439.

47. H. Huh and S.-Y. Ha, "Dynamical system approach to synchronization of the coupled Schrödinger–Lohe system", Quart. Appl. Math., **75** (2017), 555–579.

48. H. Huh, S.-Y. Ha and D. Kim, "Emergent behaviors of the Schrödinger-Lohe model on cooperative-competitive network", J. Differential Equations, **263** (2017), 8295–8321.

49. H. Huh, S.-Y. Ha and D. Kim, "Asymptotic behavior and stability for the Schrödinger-Lohe model", J. Math. Phys., **59** (2018), 102701.

50. V. Jaćimović and A. Crnkić, "Low-dimensional dynamics in non-Abelian Kuramoto model on the 3-sphere", Chaos, **28** (2018), 083105.

51. T. Kanamori and A. Takeda, "Non-convex optimization on Stiefel manifold and applications to machine learning", International Conference on Neural Information Processing, Springer, 2012, 109–116.

52. J. Kennedy and R. Eberhart, "Particle swarm optimization", Proceedings of ICNN'95-International Conference on Neural Networks, vol. 4, IEEE, 1995, 1942–1948.

53. Y. Kuramoto, *Self-entrainment of a population of coupled non-linear oscillators*, In *International Symposium on Mathematical Problems in Mathematical Physics*. Lecture Notes in Theoretical Physics **30** (1975), 420–422.

54. N. E. Leonard, D. A. Paley, F. Lekien, R. Sepulchre, D. M. Fratantoni and

R. E. Davis, "Collective motion, sensor networks and ocean sampling", Proc. IEEE, **95** (2007), 48–74.

55. M. A. Lohe, "On the double sphere model of synchronization", Phys. D., **412** (2020), 132642.

56. M. A. Lohe, "Quantum synchronization over quantum networks", J. Phys. A, **43** (2010), 465301.

57. M. A. Lohe, "Non-Abelian Kuramoto model and synchronization", J. Phys. A, **42** (2009), 395101.

58. J. Markdahl, J. Thunberg and J. Gonalves, "Almost global consensus on the n-sphere", IEEE Trans. Automat. Control, **63** (2018), 1664–1675.

59. R. Olfati-Saber, "Swarms on sphere: A programmable swarm with synchronous behaviors like oscillator network", Proc. of the 45th IEEE conference on Decision and Control, (2006), 5060–5066.

60. S. Motsch and E. Tadmor, "Heterophilious dynamics enhances consensus", SIAM. Rev., **56** (2014), 577–621.

61. D. A. Paley, N. E. Leonard, R. Sepulchre, D. Grunbaum and J. K. Parrish, "Oscillator models and collective motion", IEEE Contr. Syst. Mag., **27** (2007), 89–105.

62. C. S. Peskin, *Mathematical Aspects of Heart Physiology*, Courant Institute of Mathematical Sciences, New York, 1975.

63. A. Pikovsky, M. Rosenblum and J. Kurths, *Synchronization: A universal concept in nonlinear sciences.* Cambridge University Press, Cambridge, 2001.

64. R. Pinnau, C. Totzeck, O. Tse and S. Martin, "A consensus-based model for global optimization and its mean-field limit", Math. Models Methods Appl. Sci., **27** (2017), 183–204.

65. C. W. Reynolds, "Flocks, herds and schools: A distributed behavioral model", Proceeding SIGGRAPH 87 Proceedings of the 14th annual conference on Computer graphics and interactive techniques, (1987), 25–34.

66. S. H. Strogatz, "From Kuramoto to Crawford: Exploring the onset of synchronization in populations of coupled oscillators", Phys. D, **143** (2000), 1–20.

67. J. Thunberg, J. Markdahl, F. Bernard and J. Goncalves, "A lifting method for analyzing distributed synchronization on the unit sphere", Automatica, **96** (2018), 253–258.

68. C. M. Topaz, A. L. Bertozzi and M. A. Lewis, "A nonlocal continuum model for biological aggregation", Bull. Math. Biol., **68** (2006), 1601–1623.

69. C. M. Topaz and A. L. Bertozzi, "Swarming patterns in a two-dimensional kinematic model for biological groups", SIAM J. Appl. Math., **65** (2004), 152–174.

70. T. Vicsek, A. Czirók, E. Ben-Jacob, I. Cohen and O. Shochet, "Novel type of phase transition in a system of self-driven particles", Phys. Rev. Lett., **75** (1995), 1226–1229.

71. T. Vicsek and A. Zefeiris, "Collective motion", Phys. Rep., **517** (2012), 71–140.

72. A. T. Winfree, *The geometry of biological time.* Springer, New York, 1980.

73. A. T. Winfree, "Biological rhythms and the behavior of populations of coupled oscillators", J. Theor. Biol., **16** (1967), 15–42.
74. J. Zhu, "Synchronization of Kuramoto model in a high-dimensional linear space", Phys. Lett. A, **377** (2013), 2939–2943.

Mean-field Particle Swarm Optimization

Sara Grassi

Department of Mathematical, Physical and Computer Sciences,
University of Parma,
Parco Area delle Scienze 7/A, 43124 Parma, Italy
sara.grassi@unipr.it

Hui Huang

Department of Mathematics and Statistics,
University of Calgary,
2500 University Dr NW, Calgary, AB T2N 1N4, Canada
hui.huang1@ucalgary.ca

Lorenzo Pareschi

Department of Mathematics and Computer Science,
University of Ferrara,
Via Machiavelli 30, 44121 Ferrara, Italy
lorenzo.pareschi@unife.it

Jinniao Qiu

Department of Mathematics and Statistics,
University of Calgary,
2500 University Dr NW, Calgary, AB T2N 1N4, Canada
jinniao.qiu@ucalgary.ca

In this chapter we survey some recent results on the global minimization of a non-convex and possibly non-smooth high dimensional objective function by means of particle-based gradient-free methods. Such
problems arise in many situations of contemporary interest in machine
learning and signal processing. After a brief overview of metaheuristic methods based on particle swarm optimization (PSO), we introduce a continuous formulation via second-order systems of stochastic

differential equations that generalize PSO methods and provide the basis for their theoretical analysis. Subsequently, we will show how through the use of mean-field techniques it is possible to derive in the limit of large particles number the corresponding mean-field PSO description based on Vlasov-Fokker-Planck type equations. Finally, in the zero inertia limit, we will analyze the corresponding macroscopic hydrodynamic equations, showing that they generalize the recently introduced consensus-based optimization (CBO) methods by including memory effects. Rigorous results concerning the mean-field limit, the zero-inertia limit, and the convergence of the mean-field PSO method towards the global minimum are provided along with a suite of numerical examples.

Contents

1. Introduction

The Particle Swarm Optimization (PSO) algorithm was introduced by James Kennedy, a social psychologist, and Russel Eberhart, an electrical engineer, in the mid-1990s [1,2]. Since its introduction, due to its simplicity and versatility, the PSO method has gained a great deal of attention from the scientific community, resulting in a huge number of variants of the original algorithm [3–7]. The origin of the method can actually be traced back to an earlier time, since the basic principle of optimization by interacting agents is inspired by previous attempts to reproduce the observed behaviors of animals in their natural habitat, such as flocks of birds or schools of fish [8–13]. These roots in the natural processes of collective animal behavior lead to the PSO algorithm's classification as belonging to Swarm Intelligence (SI), where the notion of swarm intelligence refers to the property of a system in which the coordinated behaviors of agents interacting locally with their environment cause coherent global functional patterns (e.g., self-organization, emergent behavior) to emerge [14–18].

Currently, similar to other gradient-free approaches [19–25], PSO is considered an efficient metaheuristic method for solving complex optimization problems and is available in several programming language libraries. Gradient-based optimizers are effective at finding local minima for high-dimensional, nonlinearly constrained convex problems; however, most gradient-based optimizers have problems dealing with noisy, discontinuous functions, and are not designed to handle discrete and mixed discrete-continuous variables. Unlike gradient-based methods in a convex search space, metaheuristic methods are not necessarily guaranteed to find true global optimal solutions, but they are capable of finding many good solutions that are sometimes sufficient in practical applications. Some of the most popular stochastic metaheuristic methods include Simulated Annealing (SA) [26–28], Ant Colony Optimization (ACO) [29, 30], Genetic Algorithms (GA) [31, 32] and Differential Evolution (DE) [33, 34]. See also Ref. [22] for a recent survey on other natured inspired metaheuristics. It should also be mentioned that a large number of newer metaheuristic methods have begun to attract criticism in the research community for hiding their lack of novelty behind elaborate constructions unsupported by any theoretical analysis [35].

In spite of its apparent simplicity, PSO poses formidable challenges for those interested in understanding swarm intelligence through theoretical analysis. To date a fully complete mathematical theory for particle swarm optimization is still lacking (see for example Refs. [36–41] and the references therein). The algorithm explores the search space in an intelligent way thanks to a population of particles interacting with each other and updated at each step their position and velocity. Thus, from the theoretical point of view, one can take advantage of the fact that PSO is inspired by classical second order Newtonian dynamics of particle systems. This allows approaches derived from statistical mechanics and mean-field theory to be adapted to the study of the system properties in the limit of a large number of particles [42–49].

Analogies with mean-field dynamics in consensus formation have recently inspired Consensus-based Optimization (CBO) methods, a novel class of particle based methods for global optimization (see Refs. [50–58] and the recent survey [59]). Global optimization methods with similar features, but based on Kuramoto-Vicseck dynamics constrained to hypersurfaces [60–62] or on binary Boltzmann dynamics [63], have been introduced and studied recently. These methods are inherently simpler than PSO methods since they were inspired by first order consensus-like dynamics typical of social interactions such as opinion formations and wealth exchanges [64,65]. In contrast to classic metaheuristic methods typically formulated through a discrete sequence of operations and for which it is quite difficult to provide rigorous convergence to global minimizers, CBO-like methods, thanks to their formulation through stochastic differential equations (SDE) permit to exploit mean-field techniques to prove global convergence for a large class of optimization problems [51, 52, 61, 62]. On the other hand, CBO methods seem to be powerful and robust enough to tackle many interesting high dimensional non-convex optimization problems of interest in machine learning and sampling [52, 57, 61–63, 66, 67].

In this chapter we review some recent results on the mean-field modeling of particle swarm optimization with the goal of providing a robust mathematical theory for PSO methods and their convergence to the global minimum, based on a continuous description of their dynamics [67–72]. A major difficulty in the mathematical description of PSO methods, and other metaheuristic algorithms, is the presence of memory mechanisms that make their interpretation in terms of differential equations particularly challenging. To this end, the discrete PSO method is generalized via a system of second-order SDEs in which an additional state variable takes into account the memory of the individual particle. We refer to Ref. [53] for alternative approaches to memory mechanisms in CBO system.

Adopting the same regularization process for the global best as in CBO methods [50, 51], it is then possible to pass to the mean-field limit and derive the corresponding Vlasov-Fokker-Planck equation that characterizes the behavior of the system in the limit of a large number of particles [68,70]. The new mathematical formalism based on mean-field equations permits to study the behavior of the Vlasov-Fokker-Planck PSO model in the limit of zero inertia (see Refs. [73–78] for related results in other contexts). In particular, we prove that in this limit the PSO dynamics is described by simplified macroscopic models that correspond to a generalization of CBO models including memory effects and local best [68, 71]. The convergence of the mean-field PSO model to the global minimum is then discussed and shown rigorously in absence of memory effects [72].

Several numerical examples are reported to validate of the mean-field process and the small inertia limit, and to illustrate the role of the various parameters involved in solving high dimensional global optimization problems for various prototype test functions. Other than the basic algorithmic aspects of implementing these generalized PSO methods, we do not discuss the practical algorithmic improvements that can be adopted to increase the success rate, like for example the use of random batch methods [52, 79,80], particle reduction techniques [61, 62] and parameters adaptivity [4, 7]. We refer to Ref. [67] for further details on these implementation aspects.

The rest of the survey is organized as follows. In Sec. 2 we introduce the PSO algorithms and derive the corresponding representations as SDEs using a time continuous approximation of the memory process. Next, in Sec. 3, thanks to a regularization of the global best and the local best we discuss the large particle limit and derive the respective Vlasov-Fokker-Planck equations describing the mean-field dynamic. A rigorous proof of the mean-field limit is also given. Section 4 is then dedicated to the zero-inertia limit for the mean-field system that allows to recover a CBO model with local best as the corresponding macroscopic limit. This is shown rigorously at the end of the section. A general convergence result to the global minimum is illustrated in Sec. 5 in absence of memory effects. Several numerical examples, validating the mean-field approximation, the small inertia limit and testing the performances of the minimizers against some prototype functions in high dimension are then given in Sec. 6. Some concluding remarks and open research directions are reported at the end of the chapter.

2. Second order stochastic models for particle swarm optimization

In the sequel we consider the following optimization problem

$$x^* \in \arg\min_{x \in \mathbb{R}^d} \mathcal{F}(x), \qquad (2.1)$$

where $\mathcal{F}(x) : \mathbb{R}^d \to \mathbb{R}$ is a given high dimensional objective function, which we wish to minimize. In machine learning the objective function allows the algorithm designer to encode the appropriate and expected behavior for the machine learning model, such as fitting well to the training data versus some loss function. Modern applications frequently require learning algorithms to operate in extremely high dimensional spaces [81,82]. In other applications, the natural objective of the learning task is a possibly non-smooth and non-convex function [83]. Common examples include training deep neural networks and tensor decomposition problems. In contrast to gradient-based optimizers and other metaheuristic solvers, PSO solve the minimization problem (2.1) by starting from a population of candidate solutions, represented by particles, and moving these particles in the search space according to simple mathematical relationships on particle position and speed. The movement of each particle is influenced by its best known local position, but it is also driven to the best collective position of the swarm in the search space, which is updated when the particles find better positions (see Fig. 1).

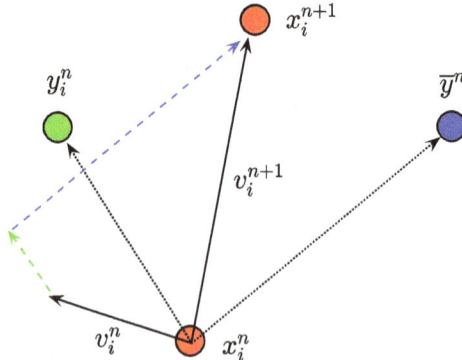

Fig. 1. Particle dynamics in the standard PSO model (2.2). Green and blue dashed arrows denote the influence of the local best and global best, respectively.

2.1. *The standard PSO algorithm*

The method is based on introducing N particles with position $x_i \in \mathbb{R}^d$ and speed $v_i \in \mathbb{R}^d$, $i = 1, \ldots, N$. In the *standard PSO algorithm* the particle positions and velocities, starting with an initial x_i^0 and v_i^0 assigned, are updated according to the following rule [1]

$$
\begin{aligned}
x_i^{n+1} &= x_i^n + v_i^{n+1}, \\
v_i^{n+1} &= v_i^n + c_1 R_1^n \left(y_i^n - x_i^n \right) + c_2 R_2^n \left(\bar{y}^n - x_i^n \right),
\end{aligned}
\tag{2.2}
$$

where the values $c_1, c_2 \in \mathbb{R}$ are the *acceleration coefficients*, y_i^n is the *local best* position found by the i particle up to that iteration, and $\bar{y}^n$ is the *global best* position found among all the particles up to that iteration. The terms R_1^n and R_2^n denote two d-dimensional diagonal matrices with random numbers uniformly distributed in $[0, 1]$ on their diagonals. These numbers are generated at each iteration and for each particle. Typically, the values of x_i and v_i are restricted within a specific search domain $X = [X_{min}, X_{max}]^d$ and velocity range $V = [-V_{max}, V_{max}]^d$. Different boundary conditions are usually applied in the search space X.

The local best y_i^n and global best $\bar{y}^n$ are defined by the following relationships

$$
\begin{aligned}
y_i^0 &= x_i^0, \\
y_i^{n+1} &= \begin{cases} y_i^n & \text{if } \mathcal{F}(x_i^{n+1}) \geq \mathcal{F}(y_i^n), \\ x_i^{n+1} & \text{if } \mathcal{F}(x_i^{n+1}) < \mathcal{F}(y_i^n), \end{cases} \\[4pt]
\bar{y}^0 &= \operatorname{argmin}\{\mathcal{F}(x_1^0), \mathcal{F}(x_2^0), \ldots, \mathcal{F}(x_N^0)\}, \\
\bar{y}^{n+1} &= \operatorname{argmin}\{\mathcal{F}(y_1^{n+1}), \mathcal{F}(y_2^{n+1}), \ldots, \mathcal{F}(y_N^{n+1})\}.
\end{aligned}
\tag{2.3}
$$

Another way to represent the local best, which will be useful in the sequel, is the following [6]

$$
y_i^{n+1} = y_i^n + \frac{1}{2} \left(x_i^{n+1} - y_i^n \right) S(x_i^{n+1}, y_i^n),
\tag{2.4}
$$

where

$$
S(x, y) = (1 + \operatorname{sgn}(\mathcal{F}(y) - \mathcal{F}(x))).
\tag{2.5}
$$

2.2. The stochastic differential PSO system

In order to derive a time continuous version of the PSO algorithm (2.2), we rewrite it in the form

$$
\begin{aligned}
x_i^{n+1} &= x_i^n + v_i^{n+1}, \\
v_i^{n+1} &= v_i^n + \frac{c_1}{2}\left(y_i^n - x_i^n\right) + \frac{c_2}{2}\left(\bar{y}^n - x_i^n\right) \\
&\quad + \frac{c_1}{2}\widetilde{R}_1\left(y_i^n - x_i^n\right) + \frac{c_2}{2}\widetilde{R}_2\left(\bar{y}^n - x_i^n\right),
\end{aligned}
\tag{2.6}
$$

where $\widetilde{R}_k = (2R_k - 1)$, $k = 1, 2$. We can interpret (2.6) as a semi-implicit time discretization method for SDEs with time stepping $\Delta t = 1$ where the implicit Euler scheme has been used for the first equation and the Euler-Maruyama method is used for the second one. Note that, the particular distribution of the random noise will not change the corresponding stochastic differential system provided the noise has the same mean value and variance. In the case of the PSO model (2.6), since the random terms are uniformly distributed in $[-1, 1]$, the mean value is 0 and the corresponding variance is $1/3$.

We can then write the time continuous formulation as a second order system of SDEs in Itô form defining the *stochastic differential PSO system*

$$
\begin{aligned}
dX_t^i &= V_t^i dt, \\
dV_t^i &= \lambda_1\left(Y_t^i - X_t^i\right) dt + \lambda_2\left(\bar{Y}_t - X_t^i\right) dt \\
&\quad + \sigma_1 D(Y_t^i - X_t^i) dB_t^{1,i} + \sigma_2 D(\bar{Y}_t - X_t^i) dB_t^{2,i},
\end{aligned}
\tag{2.7}
$$

with

$$
\lambda_k = \frac{c_k}{2}, \quad \sigma_k = \frac{c_k}{2\sqrt{3}}, \quad k = 1, 2
\tag{2.8}
$$

the *drift and diffusion coefficients* and

$$
D(X_t) = \mathrm{diag}\left\{(X_t)_1, (X_t)_2, \ldots, (X_t)_d\right\},
\tag{2.9}
$$

a d-dimensional diagonal matrix.

In (2.7) the vectors $B_t^k = \left((B_t^k)_1, (B_t^k)_2, \ldots, (B_t^k)_d\right)^T$, $k = 1, 2$ denote d independent 1-dimensional *Brownian motions* and depend on the i-th particle. One critical aspect is the definition of the best positions Y_t^i and $\bar{Y}_t$ which in the PSO method make use of the past history of the particles. Thanks to (2.4), for a positive constant ν, we can approximate y_i^{n+1} with

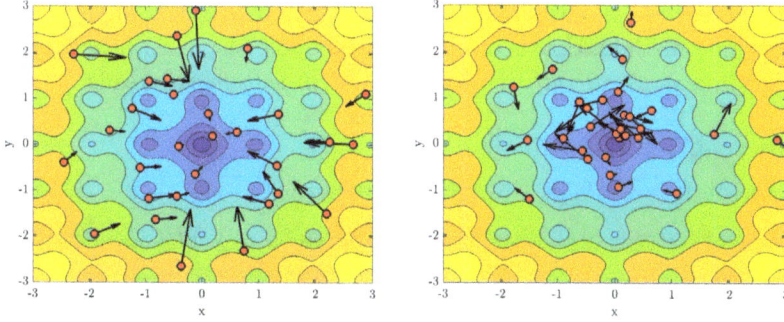

Fig. 2. Snapshots of the PSO minimization process (2.12) for the two-dimensional Ackley function (see Table 5) using $N = 30$ particles, with $m = 0$, $c_1 = 0.25$ and $c_2 = 2$.

the following *differential system for the local best*

$$dY_t^i = \nu \left(X_t^i - Y_t^i \right) S(X_t^i, Y_t^i)dt, \tag{2.10}$$

with $Y_0^i = X_0^i$ and consequently define

$$\overline{Y}_t = \operatorname{argmin} \left\{ \mathcal{F}(Y_t^1), \mathcal{F}(Y_t^2), \dots, \mathcal{F}(Y_t^N) \right\}. \tag{2.11}$$

Note that, Eq. (2.10) does not describe the evolution of the local best, but rather a time continuous approximation of its evolution.

2.3. *Stochastic differential PSO model with inertia*

To optimize the search algorithm, the value $c_k = 2$, $k = 1, 2$ was adopted in early PSO research. This value, which corresponds to $\lambda_k = 1$ and $\sigma_k = 1/\sqrt{3}$, $k = 1, 2$ in the SDEs form, however, may lead to unstable dynamics with particle speed increase without control. The use of hard bounds on velocity in $[-V_{\max}, V_{\max}]^d$ is one way to control the velocities. However, the value of $V_{\max}$ is problem-specific and difficult to determine. For this reason, the *PSO algorithm with inertia* has been considered [5]

$$\begin{aligned} x_i^{n+1} &= x_i^n + v_i^{n+1}, \\ v_i^{n+1} &= mv_i^n + c_1 R_1^n \left(y_i^n - x_i^n \right) + c_2 R_2^n \left(\overline{y}^n - x_i^n \right), \end{aligned} \tag{2.12}$$

where $m \in (0, 1]$ is the *inertia weight*. The above system can be rewritten as

$$\begin{aligned} x_i^{n+1} &= x_i^n + v_i^{n+1}, \\ mv_i^{n+1} &= mv_i^n - (1 - m)v_i^{n+1} + c_1 R_1^n \left(y_i^n - x_i^n \right) + c_2 R_2^n \left(\overline{y}^n - x_i^n \right). \end{aligned} \tag{2.13}$$

In this case, we can interpret the second equation as a semi-implicit Euler-Maruyama method, that is implicit in v_i and explicit in x_i, hence the corresponding *stochastic differential PSO system with inertia* reads

$$
\begin{aligned}
dX_t^i &= V_t^i dt, \\
mdV_t^i &= -\gamma V_t^i dt + \lambda_1 \left(Y_t^i - X_t^i \right) dt + \lambda_2 \left(\overline{Y}_t - X_t^i \right) dt \\
&\quad + \sigma_1 D(Y_t^i - X_t^i) dB_t^{1,i} + \sigma_2 D(\overline{Y}_t - X_t^i) dB_t^{2,i},
\end{aligned}
\tag{2.14}
$$

where $\gamma = (1 - m) \geq 0$ is the *friction coefficient*. Thus, the constant γ acts effectively as a friction coefficient, and can be related to the fluidity of the medium in which particles move. System (2.14) is reminiscent of other second order stochastic particle system with inertia [73,74]. However, note that here, the inertia weight m and the friction coefficient γ are not independent.

In practice, in the PSO method (2.12) the parameter γ is often initially set to some low value, which corresponds to a system where particles move in a low viscosity medium and perform extensive exploration, and gradually increased to a higher value closer to one, where the system is more dissipative and would more easily concentrate into local minima. Most PSO approaches, nowadays, are based on (2.12) (or some variant) which is usually referred to as canonical PSO method to distinguish it from the original PSO method (2.2) (see Ref. [4]). Similarly we will refer to (2.7)–(2.10) as the original stochastic differential PSO (SD-PSO) system and to (2.14)–(2.10) as the canonical SD-PSO system.

Remark 2.1: We underline that the PSO stochastic systems (2.14)–(2.10) if discretized properly yields the PSO algorithm with inertia (2.12). This is achieved discretizing (2.14) implicitly in V_t^i and explicitly in X_t^i, and (2.10) implicitly in X_t^i and explicitly in Y_t^i. Taking $\Delta t = 1$, $\nu = 1/2$, the drift and diffusion terms satisfying (2.8), and a uniform noise permits to recover exactly (2.12). We refer to the last part of the chapter containing the numerical examples for further details.

3. Mean-field particle swarm optimization

In this section we introduce a modified version of the canonical stochastic differential PSO system for which we can formally compute its mean-field limit. We first consider the case in absence of memory effects and then we extend the results to the general case. Throughout this chapter, our theoretical analysis assumes the cost function $\mathcal{F}$ satisfies the following

Assumption 1: For the given cost function $\mathcal{F} : \mathbb{R}^d \to \mathbb{R}$, it holds that:

(1) There exists some constant $L > 0$ such $|\mathcal{F}(x) - \mathcal{F}(y)| \leq L(|x| + |y|)|x - y|$ for all $x, y \in \mathbb{R}^d$;

(2) $\mathcal{F}$ is bounded from below with $-\infty < \underline{\mathcal{F}} := \inf \mathcal{F}$ and there exists some constant $C_u > 0$ such that

$$\mathcal{F}(x) - \underline{\mathcal{F}} \leq C_u(1 + |x|^2) \text{ for all } x \in \mathbb{R}^d;$$

(3) $\mathcal{F}$ has quadratic growth at infinity. Namely, there exist constants $C_l, M > 0$ such that

$$\mathcal{F}(x) - \underline{\mathcal{F}} \geq C_l |x|^2 \text{ for all } |x| \geq M.$$

3.1. *Regularized PSO dynamics without memory effects*

To simplify the mathematical description, let us consider a PSO approach where the dynamic is instantaneous without memory of the local best positions and the global best has been regularized as in Ref. [50]. The corresponding second order system of SDEs describing the *regularized SD-PSO dynamics* takes the form[a]

$$dX_t^{i,N} = V_t^{i,N} dt,$$
$$mdV_t^{i,N} = -\gamma V_t^{i,N} dt + \lambda \left(X^\alpha \left(\rho_t^N \right) - X_t^{i,N} \right) dt \tag{3.1}$$
$$+ \sigma D \left(X^\alpha \left(\rho_t^N \right) - X_t^{i,N} \right) dB_t^i,$$

where the $\mathbb{R}^d$-valued functions $X_t^{i,N}$ and $V_t^{i,N}$ denote the position and velocity of the i-th particle at time t, and $\{(B_t^i)_{t \geq 0}\}_{i=1}^N$ are N independent d-dimensional Brownian motions. Here the weighted average *regularization of the global best* is given by

$$X^\alpha(\rho_t^N) := \frac{\int_{\mathbb{R}^d} x \omega_\alpha^{\mathcal{F}}(x) \rho_t^N(dx)}{\int_{\mathbb{R}^d} \omega_\alpha^{\mathcal{F}}(x) \rho_t^N(dx)}, \tag{3.2}$$

with the empirical measure $\rho^N := \frac{1}{N} \sum_{i=1}^N \delta_{X^{i,N}}$, which is the spacial marginal of $f^N := \frac{1}{N} \sum_{i=1}^N \delta_{(X^{i,N}, V^{i,N})}$. The choice of the weight function

---

[a]The superscript N is used to emphasize the dependence on the number of particles in the system.

$\omega_\alpha^{\mathcal{F}}(x) := e^{-\alpha \mathcal{F}(x)}$ in (3.2) comes from the well-known *Laplace principle*, a classical result in large deviation theory, which states that for any probability measure $\rho \in \mathcal{P}(\mathbb{R}^d)$ compactly supported, it holds

$$\lim_{\alpha \to \infty} \left(-\frac{1}{\alpha} \log \left(\int_{\mathbb{R}^d} e^{-\alpha \mathcal{F}(x)} \rho(dx) \right) \right) = \inf_{x \in \text{supp}(\rho)} \mathcal{F}(x). \qquad (3.3)$$

Therefore, for large values of $\alpha \gg 1$ the regularized global best $X^\alpha(\rho_t^N) \approx X_t^*$, where

$$X_t^* = \text{argmin} \left\{ \mathcal{F}(X_t^{1,N}), \mathcal{F}(X_t^{2,N}), \ldots, \mathcal{F}(X_t^{N,N}) \right\}.$$

We emphasize that the stochastic particle system (3.1) has locally Lipschitz coefficients, thus it admits strong solutions and pathwise uniqueness holds up to any finite time $T > 0$, see Refs. [18, 84]. The above system of SDEs in the sequel is considered in a general setting, without necessarily satisfying the PSO constraint (2.8).

As the particle number $N \to \infty$, one expects to derive the *mean-filed PSO description without local best* characterized by the following nonlinear *Vlasov-Fokker-Planck equation*

$$\partial_t f + v \cdot \nabla_x f =$$
$$\nabla_v \cdot \left(\frac{\gamma}{m} v f + \frac{\lambda}{m} (x - X^\alpha(\rho)) f + \frac{\sigma^2}{2m^2} D(x - X^\alpha(\rho))^2 \nabla_v f \right) \qquad (3.4)$$

where we have used the identity

$$\sum_{j=1}^{d} \frac{\partial^2}{\partial v_j^2} \left((x - X^\alpha(\rho))_j^2 f \right) = \nabla_v \cdot \left(D(x - X^\alpha(\rho))^2 \nabla_v f \right) \qquad (3.5)$$

with $D(x - X^\alpha(\rho))^2$ the diagonal matrix given by the square of $D(x - X^\alpha(\rho))$. Equation (3.4) represents the mean-field PSO (MF-PSO) model without local best and should be accompanied by initial (and boundary) data, and normalization

$$\iint_{\mathbb{R}^{2d}} f(t, dx, dv) = 1.$$

We refer to Refs. [13, 43–45, 49] and the references therein, for more details and rigorous results about mean-field models of Vlasov-Fokker-Planck type. Note, however, that the presence of $X^\alpha(\rho)$ makes the Vlasov-Fokker-Planck equation non-linear and non-local. This is non-standard in the literature and raises several analytical and numerical questions (see Refs. [51, 61]).

3.1.1. *Mean-field limit*

In this section, following Ref. [70] we provide a rigorous justification of the mean-field limit of PSO model (3.1) towards its mean-field PDE (3.4) through a compactness argument. More precisely, we first prove that the sequence of empirical measures $\{f^N\}_{N\geq 2}$ ($f^N = \frac{1}{N}\sum_{i=1}^{N}\delta_{(X^{i,N},V^{i,N})}$ are $\mathcal{P}(\mathcal{C}([0,T];\mathbb{R}^d) \times \mathcal{C}([0,T];\mathbb{R}^d))$-valued random variables) is tight. Prokhorov's theorem indicates that there exists a subsequence of $\{f^N\}_{N\geq 2}$ converging in law to a random measure f. Then, to identify the limit, we verify that the limit measure f is a weak solution to the mean-field PSO equation (3.4) almost surely, while the uniqueness of the weak solution to PDE (3.4) yields that f is actually deterministic. Our main result can be described in the following way:

Theorem 1: *Let $\mathcal{F}$ satisfy Assumption 1 and $f_0 \in \mathcal{P}_4(\mathbb{R}^{2d})$. For any $N \geq 2$, we assume that $\{(X_t^{i,N}, V_t^{i,N})_{t\in[0,T]}\}_{i=1}^{N}$ is the unique solution to the SD-PSO system (3.1) with $f_0^{\otimes N}$-distributed initial data $\{X_0^{i,N}, V_0^{i,N}\}_{i=1}^{N}$. Then the limit (denoted by f) of the sequence of the empirical measure $f^N = \frac{1}{N}\sum_{i=1}^{N}\delta_{(X^{i,N},V^{i,N})}$ exists. Moreover, f is the unique weak solution to the MF-PSO equation (3.4).*

To obtain the above theorem, let us first prove the following lemma on a uniform moment estimate for the particle system (3.1).

Lemma 3.1: *Let $\mathcal{F}$ satisfy Assumption 1 and $f_0 \in \mathcal{P}_4(\mathbb{R}^{2d})$. For any $N \geq 2$, assume that $\{(X_t^{i,N}, V_t^{i,N})_{t\in[0,T]}\}_{i=1}^{N}$ is the unique solution to the SD-PSO system (3.1) with $f_0^{\otimes N}$-distributed initial data $\{(X_0^{i,N}, V_0^{i,N})\}_{i=1}^{N}$. Then there exists a constant $K > 0$ independent of N such that*

$$
\sup_{i=1,\cdots,N}\left\{\sup_{t\in[0,T]} \mathbb{E}\left[|X_t^{i,N}|^2 + |X_t^{i,N}|^4 + |V_t^{i,N}|^2 + |V_t^{i,N}|^4\right]\right\}
$$
$$
+ \sup_{t\in[0,T]} \mathbb{E}\left[|X^\alpha(\rho_t^N)|^2 + |X^\alpha(\rho_t^N)|^4\right] \leq K. \tag{3.6}
$$

The proof follows similar arguments as in Ref. [51, Lemma 3.4].

We treat $(X^{i,N}, V^{i,N}) : \Omega \mapsto \mathcal{C}([0,T];\mathbb{R}^d) \times \mathcal{C}([0,T];\mathbb{R}^d)$. Then $f^N = \sum_{i=1}^{N}\delta_{(X^{i,N},V^{i,N})} : \Omega \mapsto \mathcal{P}(\mathcal{C}([0,T];\mathbb{R}^d) \times \mathcal{C}([0,T];\mathbb{R}^d))$ is a random measure. Let us denote $\mathcal{L}(f^N) := \mathrm{Law}(f^N) \in \mathcal{P}(\mathcal{P}(\mathcal{C}([0,T];\mathbb{R}^d)\times\mathcal{C}([0,T];\mathbb{R}^d)))$. We can prove that $\{\mathcal{L}(f^N)\}_{N\geq 2}$ is tight, or we say $\{f^N\}_{N\geq 2}$ is tight, which can be done by verifying the Aldous criteria [85] as presented below:

Lemma 3.2: *Let $\{X^n\}_{n\in\mathbb{N}}$ be a sequence of random variables defined on a probability space $(\Omega, \mathcal{F}, \mathbb{P})$ and valued in $\mathcal{C}([0,T];\mathbb{R}^d)$. The sequence of probability distributions $\{\mu_{X^n}\}_{n\in\mathbb{N}}$ of $\{X^n\}_{n\in\mathbb{N}}$ is tight on $\mathcal{C}([0,T];\mathbb{R}^d)$ if the following two conditions hold.*

(Con1) For all $t \in [0,T]$, the set of distributions of X_t^n, denoted by $\{\mu_{X_t^n}\}_{n\in\mathbb{N}}$, is tight as a sequence of probability measures on $\mathbb{R}^d$.

(Con2) For all $\varepsilon > 0$, $\eta > 0$, there exist $\delta_0 > 0$ and $n_0 \in \mathbb{N}$ such that for all $n \geq n_0$ and for all discrete-valued $\sigma(X_s^n; s \in [0,T])$-stopping times β with $0 \leq \beta + \delta_0 \leq T$, it holds that

$$\sup_{\delta\in[0,\delta_0]} \mathbb{P}\left(|X_{\beta+\delta}^n - X_\beta^n| \geq \eta\right) \leq \varepsilon. \tag{3.7}$$

We can then prove

Theorem 3.3: *Let $\mathcal{F}$ satisfy Assumption 1 and $f_0 \in \mathcal{P}_4(\mathbb{R}^{2d})$. For any $N \geq 2$, we assume that $\{(X_t^{i,N}, V_t^{i,N})_{t\in[0,T]}\}_{i=1}^N$ is the unique solution to the SD-PSO system (3.1) with $f_0^{\otimes N}$-distributed initial data $\{X_0^{i,N}, V_0^{i,N}\}_{i=1}^N$. Then the sequence $\{\mathcal{L}(f^N)\}_{N\geq 2}$ is tight in $\mathcal{P}(\mathcal{P}(\mathcal{C}([0,T];\mathbb{R}^d) \times \mathcal{C}([0,T];\mathbb{R}^d)))$.*

Proof: According to Proposition 2.2 (ii) in Ref. [43], we only need to prove that $\{\mathcal{L}((X^{1,N}, V^{1,N}))\}_{N\geq 2}$ is tight in $\mathcal{P}(\mathcal{C}([0,T];\mathbb{R}^d) \times \mathcal{C}([0,T];\mathbb{R}^d))$ because of the exchangeability of the particle system. It is sufficient to justify conditions $(Con1)$ and $(Con2)$ in Lemma 3.2.

• *Step 1: Checking $(Con1)$.* For any $\varepsilon > 0$, there exists a compact subset $U_\varepsilon := \{(x,v) : |x|^2 + |v|^2 \leq \frac{K}{\varepsilon}\}$ such that by Markov's inequality

$$\mathcal{L}((X_t^{1,N}, V_t^{1,N})) \left((U_\varepsilon)^c\right) = \mathbb{P}\left(|X_t^{1,N}|^2 + |V_t^{1,N}|^2 > \frac{K}{\varepsilon}\right)$$

$$\leq \frac{\varepsilon\mathbb{E}[|X_t^{1,N}|^2 + |V_t^{1,N}|^2]}{K} \leq \varepsilon, \quad \forall N \geq 2,$$

where we have used Lemma 3.1 in the last inequality. This means that for each $t \in [0,T]$, the sequence $\{\mathcal{L}((X_t^{1,N}, V_t^{1,N}))\}_{N\geq 2}$ is tight, which verifies condition $(Con1)$ in Lemma 3.2.

• *Step 2: Checking $(Con2)$.* Let β be a $\sigma((X_s^{1,N}, V_s^{1,N}); s \in [0,T])$-stopping time with discrete values such that $\beta + \delta_0 \leq T$. It is easy to see that

$$\mathbb{E}[|X_{\beta+\delta}^{1,N} - X_\beta^{1,N}|^2] \leq \delta \int_0^T \mathbb{E}[|V_s^{1,N}|^2]ds \leq C\delta, \tag{3.8}$$

where $C > 0$ is independent of N by (3.6). Furthermore,

$$V_{\beta+\delta}^{1,N} - V_{\beta}^{1,N} = -\frac{\gamma}{m}\int_{\beta}^{\beta+\delta} V_s^{1,N}ds + \frac{\lambda}{m}\int_{\beta}^{\beta+\delta}(X^{\alpha}(\rho_s^N) - X_s^{1,N})ds$$

$$+ \frac{\sigma}{m}\int_{\beta}^{\beta+\delta} D(X^{\alpha}(\rho_s^N) - X_s^{1,N})dB_s^1 .$$

Notice that

$$\mathbb{E}\left[\left|\int_{\beta}^{\beta+\delta}(X^{\alpha}(\rho_s^N) - X_s^{1,N})ds\right|^2\right] \leq \delta\int_0^T \mathbb{E}\left[|X^{\alpha}(\rho_s^N) - X_s^{1,N}|^2\right]ds$$

$$\leq 2\delta T\left(\sup_{t\in[0,T]}\mathbb{E}\left[|X_t^{1,N}|^2\right] + \sup_{t\in[0,T]}\mathbb{E}\left[|X^{\alpha}(\rho_t^N)|^2\right]\right) \leq 2TK\delta ,$$

$$(3.9)$$

where we have used Lemma 3.1 in the last inequality. Similarly we have

$$\mathbb{E}\left[\left|\int_{\beta}^{\beta+\delta} V_s^{1,N}ds\right|^2\right] \leq TK\delta . \tag{3.10}$$

Further we apply Itô's isometry

$$\mathbb{E}\left[\left|\int_{\beta}^{\beta+\delta} D(X^{\alpha}(\rho_s^N) - X_s^{1,N})dB_s^1\right|^2\right] = \mathbb{E}\left[\int_{\beta}^{\beta+\delta}|X^{\alpha}(\rho_s^N) - X_s^{1,N}|^2ds\right]$$

$$\leq \delta^{\frac{1}{2}}\mathbb{E}\left[\left(\int_0^T|X^{\alpha}(\rho_s^N) - X_s^{1,N}|^4ds\right)^{\frac{1}{2}}\right]$$

$$\leq \delta^{\frac{1}{2}}\left(\int_0^T \mathbb{E}[|X^{\alpha}(\rho_s^N) - X_s^{1,N}|^4]ds\right)^{\frac{1}{2}} \leq \delta^{\frac{1}{2}}T^{\frac{1}{2}}(8K)^{\frac{1}{2}} . \tag{3.11}$$

Combining estimates (3.9)–(3.11) one has

$$\mathbb{E}[|V_{\beta+\delta}^{1,N} - V_{\beta}^{1,N}|^2] \leq C(\gamma.\lambda, m, \sigma, T, K)\left(\delta^{\frac{1}{2}} + \delta\right) . \tag{3.12}$$

Hence, for any $\varepsilon > 0$, $\eta > 0$, there exists some $\delta_0 > 0$ such that for all $N \geq 2$ it holds that

$$\sup_{\delta\in[0,\delta_0]}\mathbb{P}\left(|X_{\beta+\delta}^{1,N} - X_{\beta}^{1,N}|^2 + |V_{\beta+\delta}^{1,N} - V_{\beta}^{1,N}|^2 \geq \eta\right)$$

$$\leq \sup_{\delta\in[0,\delta_0]}\frac{\mathbb{E}\left[|X_{\beta+\delta}^{1,N} - X_{\beta}^{1,N}|^2 + |V_{\beta+\delta}^{1,N} - V_{\beta}^{1,N}|^2\right]}{\eta} \leq \varepsilon . \tag{3.13}$$

Hence (*Con*2) is verified. $\square$

For any $\varphi \in \mathcal{C}_c^2(\mathbb{R}^d \times \mathbb{R}^d)$, define a functional on $\mathcal{P}(\mathcal{C}([0,T];\mathbb{R}^d) \times \mathcal{C}([0,T];\mathbb{R}^d))$ as follows

$$F_\varphi(f) := \langle \varphi(\mathbf{x}_t, \mathbf{v}_t), f(d\mathbf{x}, d\mathbf{v}) \rangle - \langle \varphi(\mathbf{x}_0, \mathbf{v}_0), f(d\mathbf{x}, d\mathbf{v}) \rangle + \int_0^t \langle \mathbf{v}_s \cdot \nabla_x \varphi, f(d\mathbf{x}, d\mathbf{v}) \rangle ds$$

$$- \frac{\gamma}{m} \int_0^t \langle \mathbf{v}_s \cdot \nabla_v \varphi, f(d\mathbf{x}, d\mathbf{v}) \rangle ds + \frac{\lambda}{m} \int_0^t \langle (\mathbf{x}_s - X^\alpha(\rho_s)) \cdot \nabla_v \varphi, f(d\mathbf{x}, d\mathbf{v}) \rangle ds$$

$$- \frac{\sigma^2}{2m^2} \int_0^t \sum_{k=1}^d \left\langle (\mathbf{x}_s - X^\alpha(\rho_s))_k^2 \frac{\partial^2 \varphi}{\partial v_k^2}, f(d\mathbf{x}, d\mathbf{v}) \right\rangle ds$$

$$= \langle \varphi(x,v), f_t(dx, dv) \rangle - \langle \varphi(x,v), f_0(dx, dv) \rangle + \int_0^t \langle v \cdot \nabla_x \varphi, f_s(dx, dv) \rangle ds$$

$$- \frac{\gamma}{m} \int_0^t \langle v \cdot \nabla_v \varphi, f_s(dx, dv) \rangle ds + \frac{\lambda}{m} \int_0^t \langle (x - X^\alpha(\rho_s)) \cdot \nabla_v \varphi, f_s(dx, dv) \rangle ds$$

$$- \frac{\sigma^2}{2m^2} \int_0^t \sum_{k=1}^d \left\langle (x - X^\alpha(\rho_s))_k^2 \frac{\partial^2 \varphi}{\partial v_k^2}, f_s(dx, dv) \right\rangle ds,$$

for all $f \in \mathcal{P}(\mathcal{C}([0,T];\mathbb{R}^d) \times \mathcal{C}([0,T];\mathbb{R}^d))$ and $\mathbf{x}, \mathbf{v} \in \mathcal{C}([0,T];\mathbb{R}^d)$, where $\rho_t(x) = \int_{\mathbb{R}^d} f_t(x, dv)$.

Then we have the following estimate by the reasoning in Ref. [70, Proposition 3.2].

Lemma 3.4: *Let $\mathcal{F}$ satisfy Assumption 1 and $f_0 \in \mathcal{P}_4(\mathbb{R}^{2d})$. For any $N \geq 2$, assume that $\{(X_t^{i,N}, V_t^{i,N})_{t \in [0,T]}\}_{i=1}^N$ is the unique solution to the SD-PSO system (3.1) with $f_0^{\otimes N}$-distributed initial data $\{(X_0^{i,N}, V_0^{i,N})\}_{i=1}^N$. There exists a constant $C > 0$ depending only on $\sigma, \gamma, \lambda, m, K, T$, and $\|\nabla\varphi\|_\infty$ such that*

$$\mathbb{E}[|F_\varphi(f^N)|^2] \leq \frac{C}{N}, \tag{3.14}$$

where $f^N = \frac{1}{N} \sum_{i=1}^N \delta_{(X^{i,N}, V^{i,N})}$ is the empirical measure.

By Skorokhod's lemma (see Ref. [85, Theorem 6.7, page 70]), using Theorem 3.3 we may find a common probability space $(\Omega, \mathcal{F}, \mathbb{P})$ on which the processes $\{f^N\}_{N \in \mathbb{N}}$ converge to some process f as a random variable valued in $\mathcal{P}(\mathcal{C}([0,T];\mathbb{R}^d) \times \mathcal{C}([0,T];\mathbb{R}^d))$ almost surely. In particular, we have that for all $t \in [0,T]$ and $\phi \in C_b(\mathbb{R}^d \times \mathbb{R}^d)$,

$$\lim_{N \to \infty} |\langle \phi, f_t^N - f_t \rangle| + |X^\alpha(\rho_t^N) - X^\alpha(\rho_t)| = 0, \quad \text{a.s.} \tag{3.15}$$

Indeed, according to Assumption 1, one has $xe^{-\alpha\mathcal{F}(x)}, e^{-\alpha\mathcal{F}(x)} \in \mathcal{C}_b(\mathbb{R}^d)$, which gives

$$\lim_{N\to\infty} X^\alpha(\rho_t^N) = \lim_{N\to\infty} \frac{\langle xe^{-\alpha\mathcal{F}(x)}, \rho_t^N(dx)\rangle}{\langle e^{-\alpha\mathcal{F}(x)}, \rho_t^N(dx)\rangle} = \frac{\langle xe^{-\alpha\mathcal{F}(x)}, \rho_t(dx)\rangle}{\langle e^{-\alpha\mathcal{F}(x)}, \rho_t(dx)\rangle} = X^\alpha(\rho_t) \text{ a.s.}$$

Lemma 3.5: *[51, Lemma 3.3] Let $\mathcal{F}$ satisfy Assumption 1 and $\mu \in \mathcal{P}_2(\mathbb{R}^d)$. Then it holds that*

$$|X^\alpha(\mu)|^2 \le b_1 + b_2 \int_{\mathbb{R}^d} |x|^2 \mu(dx), \tag{3.16}$$

where b_1 and b_2 depends only on M, C_u, and C_l.

For each $A > 0$, it follows from (3.15) that

$$\mathbb{E}\left[\iint_{\mathbb{R}^{2d}} ((|x|^4 + |v|^4) \wedge A) f_t(dx, dv)\right]$$

$$= \mathbb{E}\left[\lim_{N\to\infty} \iint_{\mathbb{R}^{2d}} ((|x|^4 + |v|^4) \wedge A) f_t^N(dx, dv)\right]$$

$$\le \lim_{N\to\infty} \frac{\sum_{i=1}^N \mathbb{E}[|X_t^{i,N}|^4 + |V_t^{i,N}|^4]}{N} \le K,$$

where we have used Lemma 3.1. Letting $A \to \infty$, we have

$$\sup_{t\in[0,T]} \mathbb{E}\left[\iint_{\mathbb{R}^{2d}} (|x|^4 + |v|^4) f_t(dx, dv)\right] \le K. \tag{3.17}$$

Then Lemma 3.5 implies that

$$\mathbb{E}[|X^\alpha(\rho_t)|^4] < \infty, \tag{3.18}$$

for all $t \in [0, T]$. Furthermore, it holds that

$$\lim_{N\to\infty} \mathbb{E}\left[|\langle\phi, f_t^N - f_t\rangle|^2 + |X^\alpha(\rho_t^N) - X^\alpha(\rho_t)|^2\right] = 0, \tag{3.19}$$

which follows directly from the pointwise convergences of $\langle\phi, f_t^N - f_t\rangle$ and $X^\alpha(\rho_t^N) - X^\alpha(\rho_t)$, and the uniform estimate (3.6) in Lemma 3.1 and (3.18).

We can now prove the main result in Theorem 1:

Proof: (Theorem 1) Suppose the $\mathcal{P}(\mathcal{C}([0,T]; \mathbb{R}^d) \times \mathcal{C}([0,T]; \mathbb{R}^d))$-valued random variable f is the limit of a subsequence of the empirical measure $f^N = \frac{1}{N} \sum_{i=1}^N \delta_{(X^{i,N}, V^{i,N})}$. W.l.o.g., denote the subsequence by itself. We may continue to work on the above common probability space $(\Omega, \mathcal{F}, \mathbb{P})$ by Skorokhod's lemma where the convergence is holding almost surely (see

(3.15) for instance). We may first check that f_t is a.s. continuous in time. Indeed for any $\phi \in \mathcal{C}_b(\mathbb{R}^{2d})$ and $t_n \to t$ we may apply dominated convergence theorem

$$\iint_{\mathcal{C}([0,T];\mathbb{R}^d) \times \mathcal{C}([0,T];\mathbb{R}^d)} \phi(\mathbf{x}_{t_n}, \mathbf{v}_{t_n}) f(d\mathbf{x}, d\mathbf{v})$$

$$\to \iint_{\mathcal{C}([0,T];\mathbb{R}^d) \times \mathcal{C}([0,T];\mathbb{R}^d)} \phi(\mathbf{x}_t, \mathbf{v}_t) f(d\mathbf{x}, d\mathbf{v}) \quad \text{a.s.,}$$

which gives

$$\iint_{\mathbb{R}^{2d}} \phi(x,v) f_{t_n}(dx,dv) \to \iint_{\mathbb{R}^{2d}} \phi(x,v) f_t(dx,dv) \quad \text{a.s.}$$

For $\varphi \in \mathcal{C}_c^2(\mathbb{R}^{2d})$, using the convergence result in (3.19) one has

$$\lim_{N\to\infty} \mathbb{E}\left[|(\langle \varphi, f_t^N \rangle - \langle \varphi, f_0^N \rangle) - (\langle \varphi, f_t \rangle - \langle \varphi, f_0 \rangle)|\right] = 0. \tag{3.20}$$

Further we notice that

$$\left| \int_0^t \langle (x - X^\alpha(\rho_s^N)) \cdot \nabla_v \varphi, f_s^N \rangle ds - \int_0^t \langle (x - X^\alpha(\rho_s)) \cdot \nabla_v \varphi, f_s \rangle ds \right|$$

$$\leq \int_0^t \left| \langle (x - X^\alpha(\rho_s^N)) \cdot \nabla_v \varphi, f_s^N - f_s \rangle \right| ds$$

$$+ \int_0^t \left| \langle (X^\alpha(\rho_s) - X^\alpha(\rho_s^N)) \cdot \nabla_v \varphi, f_s \rangle \right| ds$$

$$=: \int_0^t |I_1^N(s)| ds + \int_0^t |I_2^N(s)| ds.$$

One computes

$$\mathbb{E}[|I_1^N(s)|]$$

$$\leq \mathbb{E}[|\langle x \cdot \nabla_v \varphi, f_s^N - f_s \rangle|] + \mathbb{E}[|X^\alpha(\rho_s^N) \cdot \langle \nabla_v \varphi, f_s^N - f_s \rangle|]$$

$$\leq \mathbb{E}[|\langle x \cdot \nabla_v \varphi, f_s^N - f_s \rangle|] + K^{\frac{1}{2}} (\mathbb{E}[|\langle \nabla_v \varphi, f_s^N - f_s \rangle|^2])^{\frac{1}{2}},$$

where we have used Lemma 3.1 in the second inequality. Since φ has a compact support, applying (3.19) leads to

$$\lim_{N\to\infty} \mathbb{E}\left[|I_1^N(s)|\right] = 0. \tag{3.21}$$

Moreover, the uniform boundedness of $\mathbb{E}\left[|I_1^N(s)|\right]$ follows directly from (3.17), (3.18), and the estimates in Lemma 3.1, which by the dominated convergence theorem implies

$$\lim_{N\to\infty} \int_0^t \mathbb{E}[|I_1^N(s)|] ds = 0. \tag{3.22}$$

As for I_2^N, we know that

$$|\langle (X^\alpha(\rho_s) - X^\alpha(\rho_s^N)) \cdot \nabla_v \varphi, f_s \rangle| \leq \|\nabla_v \varphi\|_\infty |X^\alpha(\rho_s) - X^\alpha(\rho_s^N)|. \quad (3.23)$$

Hence by (3.19) it yields that

$$\lim_{N \to \infty} \mathbb{E}[|I_2^N(s)|] = 0. \quad (3.24)$$

Again by the dominated convergence theorem, we have

$$\lim_{N \to \infty} \int_0^t \mathbb{E}[|I_2^N(s)|] ds = 0. \quad (3.25)$$

This combined with (3.22) leads to

$$\lim_{N \to \infty} \mathbb{E}\left[\left| \int_0^t \langle (x - X^\alpha(\rho_s^N)) \cdot \nabla_v \varphi, f_s^N \rangle ds \right.\right.$$
$$\left.\left. - \int_0^t \langle (x - X^\alpha(\rho_s)) \cdot \nabla_v \varphi, f_s \rangle ds \right| \right] = 0. \quad (3.26)$$

Similarly we split the error

$$\left| \int_0^t \left\langle (x - X^\alpha(\rho_s^N))_k^2 \frac{\partial^2}{\partial v_k{}^2} \varphi, f_s^N \right\rangle ds - \int_0^t \left\langle (x - X^\alpha(\rho_s))_k^2 \frac{\partial^2}{\partial v_k{}^2} \varphi, f_s \right\rangle ds \right|$$
$$\leq \left| \int_0^t \left\langle (x - X^\alpha(\rho_s^N))_k^2 \frac{\partial^2}{\partial v_k{}^2} \varphi, f_s^N - f_s \right\rangle ds \right|$$
$$+ \left| \int_0^t \left\langle ((x - X^\alpha(\rho_s^N))_k^2 - (x - X^\alpha(\rho_s))_k^2) \frac{\partial^2}{\partial v_k{}^2} \varphi, f_s \right\rangle ds \right|$$
$$=: \int_0^t |I_3^N(s)| ds + \int_0^t |I_4^N(s)| ds.$$

Following the same argument as for I_1^N and I_2^N, one has

$$\lim_{N \to \infty} \int_0^t \mathbb{E}[|I_3^N(s)|] ds = 0 \text{ and } \lim_{N \to \infty} \int_0^t \mathbb{E}[|I_4^N(s)|] ds = 0. \quad (3.27)$$

This implies that

$$\lim_{N \to \infty} \mathbb{E}\left[\left| \int_0^t \sum_{k=1}^d \left\langle (x - X^\alpha(\rho_s^N))_k^2 \frac{\partial^2}{\partial v_k{}^2} \varphi(x), f_s^N \right\rangle ds \right.\right.$$
$$\left.\left. - \int_0^t \sum_{k=1}^d \left\langle (x - X^\alpha(\rho_s))_k^2 \frac{\partial^2}{\partial v_k{}^2} \varphi(x), f_s \right\rangle ds \right| \right] = 0. \quad (3.28)$$

Moreover it is easy to get

$$\lim_{N\to\infty} \mathbb{E}\left[\left|\int_0^t \langle v \cdot \nabla_x \varphi, f_s^N \rangle ds - \int_0^t \langle v \cdot \nabla_x \varphi, f_s \rangle ds\right|\right] = 0 \qquad (3.29)$$

and

$$\lim_{N\to\infty} \mathbb{E}\left[\left|\int_0^t \langle v \cdot \nabla_v \varphi, f_s^N \rangle ds - \int_0^t \langle v \cdot \nabla_v \varphi, f_s \rangle ds\right|\right] = 0. \qquad (3.30)$$

Collecting estimates (3.20), (3.26), (3.28), (3.29) and (3.30) we have

$$\lim_{N\to\infty} \mathbb{E}[|F_\varphi(f^N) - F_\varphi(f)|] = 0. \qquad (3.31)$$

Then we have

$$\mathbb{E}[|F_\varphi(f)|] \leq \mathbb{E}[|F_\varphi(f^N) - F_\varphi(f)|] + \mathbb{E}[|F_\varphi(f^N)|]$$
$$\leq \mathbb{E}[|F_\varphi(f^N) - F_\varphi(f)|] + \frac{C}{\sqrt{N}} \to 0 \quad \text{as } N \to \infty,$$

where we have used Lemma 3.4 in the last inequality. This implies that

$$F_\varphi(f) = 0 \quad \text{a.s.} \qquad (3.32)$$

In other words, it holds that

$$\langle \varphi(x,v), f_t(dx,dv)\rangle - \langle \varphi(x,v), f_0(dx,dv)\rangle + \int_0^t \langle v \cdot \nabla_x \varphi, f_s(dx,dv)\rangle ds$$
$$- \frac{\gamma}{m} \int_0^t \langle v \cdot \nabla_v \varphi, f_s(dx,dv)\rangle ds + \frac{\lambda}{m} \int_0^t \langle (x - X^\alpha(\rho_s)) \cdot \nabla_v \varphi, f_s(dx,dv)\rangle ds$$
$$- \frac{\sigma^2}{2m^2} \int_0^t \sum_{k=1}^d \left\langle (x - X^\alpha(\rho_s))_k^2 \frac{\partial^2 \varphi}{\partial v_k^2}, f_s(dx,dv) \right\rangle ds = 0,$$

for any $\varphi \in \mathcal{C}_c^2(\mathbb{R}^{2d})$.

Until now we have proved that f a.s. is a weak solution to PDE (3.4). Finally combining the uniqueness of weak solution to (3.4) (see for example in Ref. [49]) and the arbitrariness of the subsequence of $\{f^N\}_{N\geq 2}$, the (deterministic) weak solution f to PDE (3.4) must be the limit of the whole sequence $\{f^N\}_{N\geq 2}$. We completed the proof. $\qquad \square$

3.2. *Regularized PSO dynamic with memory and local best*

Next, we consider the second order system of SDEs corresponding to the *regularized SD-PSO method with local best*

$$
\begin{aligned}
dX_t^{i,N} &= V_t^{i,N} dt, \\
dY_t^{i,N} &= \nu \left(X_t^{i,N} - Y_t^{i,N} \right) S^\beta \left(X_t^{i,N}, Y_t^{i,N} \right) dt, \\
m dV_t^{i,N} &= -\gamma V_t^{i,N} dt + \lambda_1 \left(Y_t^{i,N} - X_t^{i,N} \right) dt \\
&\quad + \lambda_2 \left(Y^\alpha \left(\bar{\rho}_t^N \right) - X_t^{i,N} \right) dt \\
&\quad + \sigma_1 D \left(Y_t^{i,N} - X_t^{i,N} \right) dB_t^{1,i} \\
&\quad + \sigma_2 D \left(Y^\alpha \left(\bar{\rho}_t^N \right) - X_t^{i,N} \right) dB_t^{2,i},
\end{aligned}
\tag{3.33}
$$

where, similarly to the previous case, we introduced the following *regularized global best*

$$
Y^\alpha(\bar{\rho}_t^N) := \frac{\int_{\mathbb{R}^d} y \omega_\alpha^{\mathcal{E}}(y) \bar{\rho}_t^N(dy)}{\int_{\mathbb{R}^d} \omega_\alpha^{\mathcal{E}}(y) \rho_t^N(dy)},
\tag{3.34}
$$

with the empirical measure $\bar{\rho}^N := \frac{1}{N} \sum_{i=1}^N \delta_{Y^{i,N}}$, which is the Y-marginal of $f^N = \frac{1}{N} \sum_{i=1}^N \delta_{(X^{i,N}, Y^{i,N}, V^{i,N})}$.

Furthermore, in the right-hand side of (3.33) we have replaced the $\text{sgn}(x)$ function with a *sigmoid*, for example the hyperbolic tangent $\tanh(\beta x)$ for $\beta \gg 1$, and consider

$$
S^\beta(x, y) = 1 + \tanh\left(\beta(\mathcal{F}(y) - \mathcal{F}(x))\right).
\tag{3.35}
$$

Thanks to these regularizations, also the stochastic particle system (3.33) has locally Lipschitz coefficients and therefore it admits strong solutions and pathwise uniqueness holds for any finite time $T > 0$. Even in this case, the system of SDEs (3.33) is generalized without restricting the search parameters to the PSO constraint (2.8).

In order to derive a mean-field description of system (3.33), we can follow the same arguments as in Sec. 3.1.1. The only difference is that we have an additional variable Y, which can be treated easily because of the regularity of the function S^β. Namely we can prove the tightness of the empirical measures $\{f^N\}_{N \geq 2}$ by verifying the Aldous criteria (Lemma 3.2). Then there exists a subsequence of $\{f^N\}_{N \geq 2}$ converging in law to a deterministic measure $f \in \mathcal{P}(\mathcal{C}([0,T];\mathbb{R}^d) \times \mathcal{C}([0,T];\mathbb{R}^d) \times \mathcal{C}([0,T];\mathbb{R}^d))$, which is the unique weak solution to the following *mean-field PSO system with local best* characterized by the nonlinear Vlasov-Fokker-Planck equation

$$\partial_t f + v \cdot \nabla_x f + \nabla_y \cdot \left(\nu(x-y)S^\beta(x,y)f\right) =$$
$$\nabla_v \cdot \left(\frac{\gamma}{m}vf + \frac{\lambda_1}{m}(x-y)f + \frac{\lambda_2}{m}(x-Y^\alpha(\bar{\rho}))f \right. \tag{3.36}$$
$$\left. + \left(\frac{\sigma_2^2}{2m^2}D(x-Y^\alpha(\bar{\rho}))^2 + \frac{\sigma_1^2}{2m^2}D(x-y)^2\right)\nabla_v f\right),$$

where $\bar{\rho}(t,y) = \int_{\mathbb{R}^{2d}} f(t,dx,y,dv)$.

This can be summarized in the following theorem

Theorem 2: *Let $\mathcal{F}$ satisfy Assumption 1 and $f_0 \in \mathcal{P}_4(\mathbb{R}^{3d})$. For any $N \geq 2$, we assume that $\{(X_t^{i,N}, Y_t^{i,N}, V_t^{i,N})_{t \in [0,T]}\}_{i=1}^N$ is the unique solution to the SD-PSO system (3.33) with $f_0^{\otimes N}$-distributed initial data $\{X_0^{i,N}, Y_0^{i,N}, V_0^{i,N}\}_{i=1}^N$. Then the limit (denoted by f) of the sequence of the empirical measure $f^N = \frac{1}{N}\sum_{i=1}^N \delta_{(X^{i,N},Y^{i,N},V^{i,N})}$ exists. Moreover, f is the unique weak solution to MF-PSO equation (3.36).*

---

4. Zero-inertia limit and consensus-based optimization

In this section we consider the asymptotic behavior of the previous Vlasov-Fokker-Planck equations modelling the PSO dynamic in the small inertia limit, i.e. $m \to 0$. We will derive the corresponding macroscopic equations which permit to recover and generalize the recently introduced consensus-based optimization (CBO) methods [52]. We refer to Refs. [74, 77] for a theoretical background concerning the related problem of the overdamped limit of nonlinear Vlasov-Fokker-Planck systems.

4.1. *The case without memory effects*

Let us first consider the simplified setting in absence of local best. Now we write down the so-called McKean-Vlasov process [86] underlying PSO equation (3.4), which is of the form[b]

$$d\overline{X}_t^m = \overline{V}_t^m \, dt, \tag{4.1a}$$

---

[b]We used the superscript m to emphasize its dependence on the inertia coefficient m.

$$dV^m_t = -\frac{\gamma}{m}\overline{V}^m_t\, dt + \frac{\lambda}{m}(X^\alpha(\rho^m_t) - \overline{X}^m_t)dt$$
$$+\frac{\sigma}{m}D(X^\alpha(\rho^m_t) - \overline{X}^m_t)dB_t\,,$$

(4.1b)

where

$$X^\alpha(\rho^m_t) = \frac{\int_{\mathbb{R}^d} x\omega^{\mathcal{E}}_\alpha(x)\rho^m(t, dx)}{\int_{\mathbb{R}^d} \omega^{\mathcal{E}}_\alpha(x)\rho^m(t, dx)}, \quad \rho^m(t, x) = \int_{\mathbb{R}^d} f^m(t, x, dv)\,,$$

(4.2)

and the initial data $(\overline{X}_0, \overline{V}_0)$ is the same as in (3.1). Here $f^m(t, x, v)$ is the distribution of $(\overline{X}^m_t, \overline{V}^m_t)$ at time t, which makes the set of equations (4.1) nonlinear. A direct application of the Itô-Doeblin formula yields that the law $f^m_t := f^m(t, \cdot, \cdot)$ at time t is a weak solution to (3.4).

To illustrate the limiting procedure, let us observe that for $m \to 0^+$ from Eq. (4.1b) we formally have

$$\overline{V}^0_t dt = \lambda\left(X^\alpha(\rho^0_t) - \overline{X}^0_t\right)dt + \sigma D\left(X^\alpha(\rho^0_t) - \overline{X}^0_t\right)dB_t,$$

where we used the fact that $\gamma = 1 - m \to 1$. Substituting the above identity into Eq. (4.1a) and omitting the superscripts gives the first order CBO system [52]

$$d\overline{X}_t = \lambda(X^\alpha(\rho_t) - \overline{X}_t)dt + \sigma D(X^\alpha(\rho_t) - \overline{X}_t)dB_t\,.$$

(4.3)

Therefore, the CBO models based on a multiplicative noise can be understood as reduced order approximations of SD-PSO dynamics.

4.1.1. *Formal derivation in the mean-field case*

In the sequel we will develop these arguments in the case of the nonlinear Vlasov-Fokker-Planck equation (3.4) describing the evolution of the distribution of (4.1). We rewrite the scaled Vlasov-Fokker-Planck system in the form

$$\partial_t f + v \cdot \nabla_x f + \frac{1}{m}\nabla_v \cdot (mvf + \lambda(X^\alpha(\rho) - x)f) = L_m(f)$$

(4.4)

where we used the fact that $\gamma = 1 - m$ and define

$$L_m(f) = \frac{1}{m}\nabla_v \cdot \left(vf + \frac{\sigma^2}{2m}D(x - X^\alpha(\rho))^2\nabla_v f\right)$$
$$= \frac{1}{m}\sum_{j=1}^{d}\frac{\sigma^2}{2}(x_j - X^\alpha_j(\rho))^2\frac{\partial}{\partial v_j}\left(\frac{2fv_j}{\sigma^2(x_j - X^\alpha_j(\rho))^2} + \frac{1}{m}\frac{\partial f}{\partial v_j}\right).$$

Note that the last equality is a direct consequence of identity (3.5). Let us now introduce the local Maxwellian with unitary mass and zero momentum

$$\mathcal{M}_m(x,v,t) = \prod_{j=1}^{d} M_m(x_j,v_j,t),$$

$$M_m(x_j,v_j,t) = \frac{m^{1/2}}{\pi^{1/2}\sigma|x_j - X_j^{\alpha}(\rho)|} \exp\left\{-\frac{mv_j^2}{\sigma^2(x_j - X_j^{\alpha}(\rho))^2}\right\},$$

then we have

$$L_m(f) = \frac{1}{m^2} \sum_{j=1}^{d} \frac{\sigma^2}{2}(x_j - X_j^{\alpha}(\rho))^2 \frac{\partial}{\partial v_j}\left(f\frac{\partial}{\partial v_j}\log\left(\frac{f}{M_m(x_j,v_j,t)}\right)\right).$$

Therefore $L_m(f)$ is of order $1/m^2$ and we can write for small values of $m \ll 1$

$$f(x,v,t) = \rho(x,t)\mathcal{M}_m(x,v,t). \tag{4.5}$$

Let us now integrate Eq. (4.4) with respect to v, and multiply the same equation by v and ingrate again with respect to v, we get

$$\frac{\partial\rho}{\partial t} + \nabla_x \cdot (\rho u) = 0$$

$$\frac{\partial\rho u}{\partial t} + \int_{\mathbb{R}^d} v\,(v \cdot \nabla_x f)\,dv = -\frac{\gamma}{m}\rho u + \frac{1}{m}\lambda(X^{\alpha}(\rho) - x)\rho$$

where

$$\rho u = \int_{\mathbb{R}^d} f(x,v,t)v\,dv.$$

Now assuming (4.5) we can compute for $m \ll 1$ the j-th component of the second term in the right-hand side of last equation as

$$\int_{\mathbb{R}^d} v_j(v \cdot \nabla_x(\rho(x,t)\mathcal{M}_m(x,v,t)))\,dv = \sum_{j=1}^{d} \frac{\partial}{\partial x_j}\left(\rho(x,t)\int_{\mathbb{R}^d} v_j(v_j\mathcal{M}_m(x,v,t))\,dv\right)$$

$$= \frac{\partial}{\partial x_j}\left(\rho(x,t)\int_{\mathbb{R}} v_j^2 M_m(x_j,v_j,t)\,dv_j\right)$$

$$= \frac{\sigma^2}{2m}\frac{\partial}{\partial x_j}\left(\rho(x,t)(x_j - X_j^{\alpha}(\rho))^2\right)$$

which provides the *macroscopic PSO system without local best*

$$\frac{\partial \rho}{\partial t} + \nabla_x \cdot (\rho u) = 0,$$

$$\frac{\partial (\rho u)_j}{\partial t} + \frac{\sigma^2}{2m} \frac{\partial}{\partial x_j} \left(\rho(x,t)(x_j - X_j^\alpha(\rho))^2 \right) \tag{4.6}$$

$$= -\frac{1-m}{m}(\rho u)_j + \frac{1}{m}\lambda(X_j^\alpha(\rho) - x_j)\rho.$$

Formally, as $m \to 0^+$, from the second equation in (4.6) we get

$$(\rho u)_j = \lambda(X_j^\alpha(\rho) - x_j)\rho - \frac{\sigma^2}{2} \frac{\partial}{\partial x_j} \left(\rho(x,t)(x_j - X_j^\alpha(\rho))^2 \right),$$

which substituted in the first equation yields the *mean-field CBO system* [52]

$$\frac{\partial \rho}{\partial t} + \nabla_x \cdot \lambda(X^\alpha(\rho) - x)\rho = \frac{\sigma^2}{2} \sum_{j=1}^{d} \frac{\partial^2}{\partial x_j^2} \left(\rho(x,t)(x_j - X_j^\alpha(\rho))^2 \right). \tag{4.7}$$

Therefore, in the small inertia limit we expect the macroscopic density in the PSO system (3.4) to be well approximated by the solution of the CBO Eq. (4.7). We emphasize that system (4.6) represents a novel mean-field optimization model with an intermediate level of description between the mean-field PSO system (4.4) and the mean-field CBO system (4.7).

4.1.2. *Rigorous derivation*

In this section, we present a rigorous derivation of the zero-inertia limit [71]. More precisely we prove that as $m \to 0^+$, the processes $\{\overline{X}^m\}$ satisfying the SDEs (4.1) converge weakly to the solution $\overline{X}$ to the SDE (4.3) in the continuous path space $\mathcal{C}([0,T]; \mathbb{R}^d)$, and a convergence rate is obtained. The main theorem can be stated as below:

Theorem 3: *Let Assumption 1 hold and $(X_t^m, V_t^m)_{t \in [0,T]}$ satisfy the system (4.1). Then as $m \to 0^+$, the sequence of stochastic processes $\{\overline{X}^m\}_{0 < m \le \frac{1}{2}}$ converge weakly to $\overline{X}$, which is the unique solution to the* *following SDE:*

$$\overline{X}_t = \overline{X}_0 + \lambda \int_0^t (X^\alpha(\rho_s) - \overline{X}_s)ds$$

$$+ \sigma \int_0^t D(X^\alpha(\rho_s) - \overline{X}_s)dB_s.$$

(4.8)

Moreover it holds that

$$\sup_{t\in[0,T]} \mathbb{E}[|\overline{X}_t^m - \overline{X}_t|^2] \le C\,m,$$

(4.9)

where the constant C *depends only on* $\mathbb{E}[|\overline{X}_0|^4]$, $\mathbb{E}[|\overline{V}_0|^4]$, M, C_u, C_l, λ, σ, d, *and* T.

Remark 4.1: It follows from the definition of Wasserstein distance that

$$\sup_{t\in[0,T]} W_2^2(\rho_t^m, \rho_t) \le \sup_{t\in[0,T]} \mathbb{E}[|\overline{X}_t^m - \overline{X}_t|^2] \le C\,m,$$

(4.10)

which in a way is consistent with the result obtained in Ref. [77, Theorem 1.3], where the authors obtained a quantified overdamped limit (with the same rate m) of the singular Vlasov-Poisson-Fokker-Planck system to the aggregation-diffusion equation.

The following theorem gives the well-posedness of the mean-field PSO dynamic (4.1) whose proof is analogous to [69, Theorem 2.3] or [51, Theorem 3.1], and thus omitted.

Theorem 4.2: *Let Assumption 1 hold. If* $(\overline{X}_0^m, \overline{V}_0^m) = (\overline{X}_0, \overline{V}_0)$ *is distributed according to* f_0 *with* $f_0 \in \mathcal{P}_4(\mathbb{R}^{2d})$, *then for each* $T > 0$ *and* $m \in (0,1]$, *the nonlinear SDE (4.1) admits a unique strong solution up to time* T *with the initial data* $(\overline{X}_0^m, \overline{V}_0^m)$ *and it holds further that*

$$\sup_{t\in[0,T]} \mathbb{E}\left[|\overline{X}_t^m|^4 + |\overline{V}_t^m|^4\right] \le e^{CT} \cdot \mathbb{E}\left[|\overline{X}_0|^4 + |\overline{V}_0|^4\right],$$

(4.11)

where C *depends only on* λ, m, σ, M, C_u, *and* C_l.

Solving (4.1b) for $\overline{V}_t^m$ gives

$$\overline{V}_t^m = e^{-\frac{\gamma}{m}t}\overline{V}_0 + \frac{\lambda}{m} \int_0^t e^{-\frac{\gamma}{m}(t-s)}(X^\alpha(\rho_s^m) - \overline{X}_s^m)ds$$

$$+ \frac{\sigma}{m} \int_0^t e^{-\frac{\gamma}{m}(t-s)} D(X^\alpha(\rho_s^m) - \overline{X}_s^m)dB_s,$$

which implies that

$$\overline{X}_t^m = \overline{X}_0 + \int_0^t \overline{V}_\tau d\tau = \overline{X}_0 + \int_0^t e^{-\frac{\gamma}{m}\tau}\overline{V}_0 d\tau$$

$$+ \frac{\lambda}{m}\int_0^t\int_0^\tau e^{-\frac{\gamma}{m}(\tau-s)}(X^\alpha(\rho_s^m) - \overline{X}_s^m)dsd\tau \qquad (4.12)$$

$$+ \frac{\sigma}{m}\int_0^t\int_0^\tau e^{-\frac{\gamma}{m}(\tau-s)}D(X^\alpha(\rho_s^m) - \overline{X}_s^m)dB_s d\tau .$$

Then $\overline{X}_t^m$ has the law ρ_t^m for each $t \geq 0$.

Each continuous stochastic process $\overline{X}^m$ may be seen as a $\mathcal{C}([0,T];\mathbb{R}^d)$-valued random function and it induces a probability measure (or law, denoted by ρ^m) on $\mathcal{C}([0,T];\mathbb{R}^d)$. We shall use the weak convergence in the space of probability measures on $\mathcal{C}([0,T];\mathbb{R}^d)$. In what follows, we write $\overline{X}^m \rightharpoonup \overline{X}$ or $\rho^m \rightharpoonup \rho$ with ρ being the law of $\overline{X}$, if $\{\rho^m\}_{m>0}$, as a sequence of probability measures, converges weakly to ρ, i.e., for each bounded continuous functional Φ on $\mathcal{C}([0,T];\mathbb{R}^d)$, there holds $\lim_{m\to 0^+}\mathbb{E}\left[\Phi(\overline{X}^m)\right] = \mathbb{E}\left[\Phi(\overline{X})\right]$. The weak convergence $\overline{X}^m \rightharpoonup \overline{X}$ is stronger than and actually implies the convergence of $\{\rho_t^m\}_{m>0}$ to ρ_t with ρ_t being the law of $\overline{X}_t$ for each $t \geq 0$, while the converse need not hold. Moreover, due to the separability and completeness of the space $\mathcal{C}([0,T];\mathbb{R}^d)$, Prohorov's theorem implies that the relative compactness is equivalent to the tightness; see Ref. [85] for more details.

Theorem 4.3: *Let Assumption 1 hold and $(X_t^m, V_t^m)_{t\in[0,T]}$ satisfy the system (4.1). For each countable subsequence $\{m_k\}_{k\in\mathbb{N}} \subset [0,\frac{1}{2}]$ with $\lim_{k\to\infty} m_k = 0$, the sequence of probability distributions $\{\rho^{m_k}\}_{k\in\mathbb{N}}$ of $\{\overline{X}^{m_k}\}_{k\in\mathbb{N}}$ is tight.*

Proof: By Lemma 3.2, it is sufficient to justify conditions (*Con*1) and (*Con*2) in Aldous tightness criteria.

- *Step 1: Checking* (*Con*1). First, for $0 < m \leq \frac{1}{2}$, recalling (4.12), we have by Fubini's theorem (see Ref. [87, Theorem 4.33] for the stochastic version)

$$\overline{X}_t^m = \overline{X}_0 + \int_0^t e^{-\frac{\gamma}{m}\tau}\overline{V}_0 d\tau + \frac{\lambda}{m}\int_0^t\int_0^\tau e^{-\frac{\gamma}{m}(\tau-s)}(X^\alpha(\rho_s^m) - \overline{X}_s^m)dsd\tau$$

$$+ \frac{\sigma}{m}\int_0^t\int_0^\tau e^{-\frac{\gamma}{m}(\tau-s)}D(X^\alpha(\rho_s^m) - \overline{X}_s^m)dB_s d\tau$$

$$= \overline{X}_0 + \int_0^t e^{-\frac{\gamma}{m}\tau}\overline{V}_0 d\tau + \frac{\lambda}{m}\int_0^t\int_s^t e^{-\frac{\gamma}{m}(\tau-s)}d\tau(X^\alpha(\rho_s^m) - \overline{X}_s^m)ds$$

$$
+ \frac{\sigma}{m} \int_0^t \int_s^t e^{-\frac{\gamma}{m}(\tau - s)} d\tau D(X^\alpha(\rho_s^m) - \overline{X}_s^m) dB_s
$$

$$
= \overline{X}_0 + \frac{m}{\gamma}(1 - e^{-\frac{\gamma}{m}t})\overline{V}_0 + \frac{\lambda}{\gamma} \int_0^t (1 - e^{-\frac{\gamma}{m}(t-s)})(X^\alpha(\rho_s^m) - \overline{X}_s^m) ds
$$

$$
+ \frac{\sigma}{\gamma} \int_0^t (1 - e^{-\frac{\gamma}{m}(t-s)}) D(X^\alpha(\rho_s^m) - \overline{X}_s^m) dB_s .
$$

(4.13)

Note here the assumption on $0 < m \le \frac{1}{2}$ ensures that $\gamma = 1 - m \in [\frac{1}{2}, 1)$, so $\frac{1}{\gamma}$ is well-defined. It follows from Hölder's inequality that

$$
|\overline{X}_t^m|^4 \le 64|\overline{X}_0|^4 + \frac{64m^4}{\gamma^4}|\overline{V}_0|^4 + \frac{64\lambda^4 t^3}{\gamma^4} \int_0^t |X^\alpha(\rho_s^m) - \overline{X}_s^m|^4 ds
$$

$$
+ \frac{64\sigma^4}{\gamma^4} \left| \int_0^t (1 - e^{-\frac{\gamma}{m}(t-s)}) D(X^\alpha(\rho_s^m) - \overline{X}_s^m) dB_s \right|^4 , \quad (4.14)
$$

where we have used the fact that for any sequence $\{a_i\}_{i=1}^n \ge 0$ and $p \ge 2$, there holds

$$
\left(\sum_{i=1}^n a_i \right)^p \le n^{p-1} \sum_{i=1}^n a_i^p .
$$

Using the moment inequality for stochastic integrals as in Ref. [88, Theorem 7.1] yields that

$$
\mathbb{E}\left[\left| \int_0^t (1 - e^{-\frac{\gamma}{m}(t-s)}) D(X^\alpha(\rho_s^m) - \overline{X}_s^m) dB_s \right|^4 \right]
$$

$$
\le d^3 \mathbb{E}\left[\sum_{k=1}^d \left| \int_0^t (1 - e^{-\frac{\gamma}{m}(t-s)}) (X^\alpha(\rho_s^m) - \overline{X}_s^m)_k dB_s^k e_k \right|^4 \right]
$$

$$
\le 36d^3 t \int_0^t \mathbb{E}\left[\sum_{k=1}^d |(X^\alpha(\rho_s^m) - \overline{X}_s^m)_k|^4 \right] ds
$$

$$
\le 36d^3 t \int_0^t \mathbb{E}\left[|X^\alpha(\rho_s^m) - \overline{X}_s^m|^4 \right] ds .
$$

Thus,

$$
\mathbb{E}[|\overline{X}_t^m|^4] \le 64\mathbb{E}[|\overline{X}_0|^4] + \frac{64m^4}{\gamma^4}\mathbb{E}[|\overline{V}_0|^4]
$$

$$
+ \frac{64(\lambda^4 t^3 + 36d^3 t\sigma^4)}{\gamma^4} \int_0^t \mathbb{E}[|X^\alpha(\rho_s^m) - \overline{X}_s^m|^4] ds .
$$

Notice that

$$\mathbb{E}[|X^\alpha(\rho_t^m) - \overline{X}_t^m|^4] \leq 8|X^\alpha(\rho_t^m)|^4 + 8\mathbb{E}[|\overline{X}_t^m|^4]$$
$$\leq 8(b_1 + b_2\mathbb{E}[|\overline{X}_t^m|^2])^2 + 8\mathbb{E}[|\overline{X}_t^m|^4] \qquad (4.15)$$
$$\leq c_1 + c_2\mathbb{E}[|\overline{X}_t^m|^4],$$

where we have used Lemma 3.5 in the second inequality, and c_1, c_2 depend only on C_u, M and C_l. Thus we have

$$\mathbb{E}[|\overline{X}_t^m|^4] \leq 64\mathbb{E}[|\overline{X}_0|^4] + \frac{64m^4}{\gamma^4}\mathbb{E}[|\overline{V}_0|^4] + c_3$$
$$+ \frac{64c_2(\lambda^4 t^3 + 36d^3 t\sigma^4)}{\gamma^4} \int_0^t \mathbb{E}[|\overline{X}_s^m|^4] ds.$$

Using Gronwall's inequality leads to

$$\mathbb{E}[|\overline{X}_t^m|^4] \leq \left(64\mathbb{E}[|\overline{X}_0|^4] + \frac{64m^4}{\gamma^4}\mathbb{E}[|\overline{V}_0|^4] + c_3 \right) \cdot$$
$$\cdot \exp\left(\frac{64c_2(\lambda^4 T^3 + 36d^3 T\sigma^4)}{\gamma^4} T \right), \qquad (4.16)$$

for all $t \in [0, T]$. Recalling $0 \leq m \leq \frac{1}{2}$ and $\frac{1}{\gamma} = \frac{1}{1-m} \leq 2$, from estimate (4.16) we obtain the boundedness:

$$\mathbb{E}[|\overline{X}_t^m|^4] \leq C(\mathbb{E}[|\overline{X}_0|^4], \mathbb{E}[|\overline{V}_0|^4], M, C_u, C_l, \lambda, d, \sigma, T). \qquad (4.17)$$

This yields that

$$\sup_{m\in(0,1]} \sup_{t\in[0,T]} \mathbb{E}[|\overline{X}_t^m|^4]$$
$$\leq C(\mathbb{E}[|\overline{X}_0|^4], \mathbb{E}[|\overline{V}_0|^4], M, C_u, C_l, \lambda, \sigma, d, T) =: C_1 \qquad (4.18)$$

where the constant $C_1 > 0$ is independent of m. Therefore, for any $\varepsilon > 0$, there exists a compact subset $K_\varepsilon := \{x : |x|^4 \leq \frac{C_1}{\varepsilon}\}$ such that by Markov's inequality

$$\rho_t^m((K_\varepsilon)^c) = \mathbb{P}(|X_t^m|^4 > \frac{C_1}{\varepsilon}) \leq \frac{\varepsilon\mathbb{E}[|X_t^m|^4]}{C_1} \leq \varepsilon, \quad \forall\, 0 < m \leq 1. \qquad (4.19)$$

This means that for each $t \in [0, T]$, each countable subset of $\{\rho_t^m\}_{0<m\leq1}$ is tight, which verifies condition $(Con1)$ in Lemma 3.2.

• *Step 2: Checking $(Con2)$.* Let β be a $\sigma(X_s^m; s \in [0, T])$-stopping time with discrete values such that $\beta + \delta_0 \leq T$. Without any loss of generality, we may assume that the concerned countable subsequence $\{m_k\}_{k\in\mathbb{N}} \subset [0, 1]$ satisfies

$m_k \leq \frac{1}{2}$ for all $k \in \mathbb{N}$; thus, we may just consider the case of $0 < m \leq \frac{1}{2}$ which indicates $\frac{1}{2} \leq \gamma < 1$. Recall (4.12) and compute

$$
\overline{X}^m_{\beta+\delta} - \overline{X}^m_\beta = \int_\beta^{\beta+\delta} \overline{V}_\tau d\tau = \int_\beta^{\beta+\delta} e^{-\frac{\gamma}{m}\tau} \overline{V}_0 d\tau
$$
$$
+ \frac{\lambda}{m} \int_\beta^{\beta+\delta} \int_0^\tau e^{-\frac{\gamma}{m}(\tau-s)}(X^\alpha(\rho_s^m) - \overline{X}^m_s)ds d\tau
$$
$$
+ \frac{\sigma}{m} \int_\beta^{\beta+\delta} \int_0^\tau e^{-\frac{\gamma}{m}(\tau-s)}D(X^\alpha(\rho_s^m) - \overline{X}^m_s)dB_s d\tau
$$

$$
= \int_\beta^{\beta+\delta} e^{-\frac{\gamma}{m}\tau} \overline{V}_0 d\tau
$$
$$
+ \frac{\lambda}{m} \int_0^\beta \int_\beta^{\beta+\delta} e^{-\frac{\gamma}{m}(\tau-s)}d\tau (X^\alpha(\rho_s^m) - \overline{X}^m_s)ds
$$
$$
+ \frac{\lambda}{m} \int_\beta^{\beta+\delta} \int_s^{\beta+\delta} e^{-\frac{\gamma}{m}(\tau-s)}d\tau (X^\alpha(\rho_s^m) - \overline{X}^m_s)ds
$$
$$
+ \frac{\sigma}{m} \int_0^\beta \int_\beta^{\beta+\delta} e^{-\frac{\gamma}{m}(\tau-s)}d\tau D(X^\alpha(\rho_s^m) - \overline{X}^m_s)dB_s
$$
$$
+ \frac{\sigma}{m} \int_\beta^{\beta+\delta} \int_s^{\beta+\delta} e^{-\frac{\gamma}{m}(\tau-s)}d\tau D(X^\alpha(\rho_s^m) - \overline{X}^m_s)dB_s .
$$

Then it yields

$$
\overline{X}^m_{\beta+\delta} - \overline{X}^m_\beta = \frac{m}{\gamma}(e^{-\frac{\gamma}{m}\beta} - e^{-\frac{\gamma}{m}(\beta+\delta)})\overline{V}_0
$$
$$
+ \frac{\lambda}{\gamma} \int_0^\beta (e^{-\frac{\gamma}{m}(\beta-s)} - e^{-\frac{\gamma}{m}(\beta+\delta-s)})(X^\alpha(\rho_s^m) - \overline{X}^m_s)ds
$$
$$
+ \frac{\lambda}{\gamma} \int_\beta^{\beta+\delta} (1 - e^{-\frac{\gamma}{m}(\beta+\delta-s)})(X^\alpha(\rho_s^m) - \overline{X}^m_s)ds \qquad (4.20)
$$
$$
+ \frac{\sigma}{\gamma} \int_0^\beta (e^{-\frac{\gamma}{m}(\beta-s)} - e^{-\frac{\gamma}{m}(\beta+\delta-s)})D(X^\alpha(\rho_s^m) - \overline{X}^m_s)dB_s
$$
$$
+ \frac{\sigma}{\gamma} \int_\beta^{\beta+\delta} (1 - e^{-\frac{\gamma}{m}(\beta+\delta-s)})D(X^\alpha(\rho_s^m) - \overline{X}^m_s)dB_s .
$$

Note that there holds $|e^{-x} - e^{-y}| \leq |x - y| \wedge 1$ for all $x, y \in [0, \infty)$. Basic computations further indicate that for each $q \geq 1$ and $\tau \in [0, T]$,

$$
\int_0^\tau \left| e^{-\frac{\gamma(\tau-s)}{m}} - e^{-\frac{\gamma(\tau+\delta-s)}{m}} \right|^q ds \leq \int_0^\tau \left(e^{-\frac{\gamma(\tau-s)}{m}} - e^{-\frac{\gamma(\tau+\delta-s)}{m}} \right) ds
$$

$$= \frac{m}{\gamma}\left(1 - e^{-\frac{\gamma\delta}{m}}\right) - \frac{m}{\gamma}\left(e^{-\frac{\gamma\tau}{m}} - e^{-\frac{\gamma(\tau+\delta)}{m}}\right)$$

$$\leq \frac{m}{\gamma} \cdot \frac{\gamma\delta}{m} = \delta,$$

and in particular,

$$\int_{\beta}^{\beta+\delta}\left(1 - e^{-\frac{\gamma(\beta+\delta-s)}{m}}\right)^q ds \leq \int_{\beta}^{\beta+\delta} 1 \, ds = \delta.$$

Then, it is obvious that

$$\mathbb{E}\left[\left|\frac{m}{\gamma}(e^{-\frac{\gamma}{m}\beta} - e^{-\frac{\gamma}{m}(\beta+\delta)})\overline{V}_0\right|^2\right] \leq \frac{m^2}{\gamma^2} \cdot \frac{\gamma^2\delta^2}{m^2}\left(\mathbb{E}[|\overline{V}_0|^4]\right)^{\frac{1}{2}} \leq \delta^2\left(\mathbb{E}[|\overline{V}_0|^4]\right)^{\frac{1}{2}}.$$

Next, it follows that

$$\mathbb{E}\left[\left|\int_0^{\beta}(e^{-\frac{\gamma}{m}(\beta-s)} - e^{-\frac{\gamma}{m}(\beta+\delta-s)})(X^\alpha(\rho_s^m) - \overline{X}_s^m)ds\right|^2\right]$$

$$\leq \mathbb{E}\left[\int_0^{\beta}|e^{-\frac{\gamma}{m}(\beta-s)} - e^{-\frac{\gamma}{m}(\beta+\delta-s)}|^2 ds \cdot \int_0^{\beta}|X^\alpha(\rho_s^m) - \overline{X}_s^m|^2 ds\right]$$

$$\leq \delta \cdot T \sup_{s\in[0,T]}\left(\mathbb{E}\left[|X^\alpha(\rho_s^m) - \overline{X}_s^m|^4\right]\right)^{1/2},$$

and analogously,

$$\mathbb{E}\left[\left|\int_{\beta}^{\beta+\delta}(1 - e^{-\frac{\gamma}{m}(\beta+\delta-s)})(X^\alpha(\rho_s^m) - \overline{X}_s^m)ds\right|^2\right]$$

$$\leq \mathbb{E}\left[\int_{\beta}^{\beta+\delta}\left(1 - e^{-\frac{\gamma(\beta+\delta-s)}{m}}\right)^2 ds \cdot \int_{\beta}^{\beta+\delta}|X^\alpha(\rho_s^m) - \overline{X}_s^m|^2 ds\right]$$

$$\leq \delta \cdot \mathbb{E}\left[\int_{\beta}^{\beta+\delta}|X^\alpha(\rho_s^m) - \overline{X}_s^m|^2 ds\right]$$

$$\leq \delta \cdot T \sup_{s\in[0,T]}\left(\mathbb{E}\left[|X^\alpha(\rho_s^m) - \overline{X}_s^m|^4\right]\right)^{1/2}.$$

Further, applying Itô's isometry gives

$$\mathbb{E}\left[\left|\int_0^{\beta}(e^{-\frac{\gamma}{m}(\beta-s)} - e^{-\frac{\gamma}{m}(\beta+\delta-s)})D(X^\alpha(\rho_s^m) - \overline{X}_s^m)dB_s\right|^2\right]$$

$$\leq d\mathbb{E}\left[\int_0^{\beta}|e^{-\frac{\gamma}{m}(\beta-s)} - e^{-\frac{\gamma}{m}(\beta+\delta-s)}|^2|X^\alpha(\rho_s^m) - \overline{X}_s^m|^2 ds\right]$$

$$\leq d \left(\mathbb{E} \left[\int_0^\beta |e^{-\frac{\gamma}{m}(\beta-s)} - e^{-\frac{\gamma}{m}(\beta+\delta-s)}|^4 ds \right] \right)^{1/2} \cdot$$

$$\cdot \left(\mathbb{E} \left[\int_0^\beta |X^\alpha(\rho_s^m) - \overline{X}_s^m|^4 ds \right] \right)^{1/2}$$

$$\leq d\delta^{1/2} \left(T \sup_{s \in [0,T]} \mathbb{E} \left[|X^\alpha(\rho_s^m) - \overline{X}_s^m|^4 \right] \right)^{1/2},$$

and analogously,

$$\mathbb{E} \left[\left| \int_\beta^{\beta+\delta} (1 - e^{-\frac{\gamma}{m}(\beta+\delta-s)}) D(X^\alpha(\rho_s^m) - \overline{X}_s^m) dB_s \right|^2 \right]$$

$$\leq d\delta^{1/2} \left(T \sup_{s \in [0,T]} \mathbb{E} \left[|X^\alpha(\rho_s^m) - \overline{X}_s^m|^4 \right] \right)^{1/2}.$$

Therefore, summing up the above estimates and recalling $0 < m \leq m_0 = \frac{1}{2}$, $\frac{1}{\gamma} \leq 2$, and the relations (4.15) and (4.18), we arrive at

$$\mathbb{E}[|\overline{X}_{\beta+\delta}^m - \overline{X}_\beta^m|^2]$$

$$\leq \frac{5}{\gamma^2} \delta^2 (\mathbb{E}[|\overline{V}_0|^4])^{\frac{1}{2}} + \frac{10}{\gamma^2} \left(\lambda^2 \delta T + \sigma^2 d (\delta T)^{1/2} \right) \sup_{s \in [0,T]} \left(\mathbb{E} \left[|X^\alpha(\rho_s^m) - \overline{X}_s^m|^4 \right] \right)^{1/2}$$

$$\leq C \left(\mathbb{E}[|\overline{X}_0|^4], \mathbb{E}[|\overline{V}_0|^4], M, C_u, C_l, \lambda, \sigma, d, T \right) \left(\delta^{\frac{1}{2}} + \delta + \delta^2 \right).$$

Hence, for any $\varepsilon > 0$, $\eta > 0$, there exists some $\delta_0 > 0$ such that for all $0 < m \leq \frac{1}{2}$ it holds that

$$\sup_{\delta \in [0,\delta_0]} \mathbb{P}(|\overline{X}_{\beta+\delta}^m - \overline{X}_\beta^m|^2 \geq \eta) \leq \sup_{\delta \in [0,\delta_0]} \frac{\mathbb{E}[|\overline{X}_{\beta+\delta}^m - \overline{X}_\beta^m|^2]}{\eta} \leq \varepsilon. \quad (4.21)$$

This justifies condition $Con2$ in Lemma 3.2. □

Next we shall identify the limit process, before which we recall a lemma on the stability estimate of the nonlinear term $X^\alpha(\rho)$.

Lemma 4.4: *[51, Lemma 3.2] Assume that $\rho, \widehat{\rho} \in \mathcal{P}_4(\mathbb{R}^d)$. Then the following stability estimate holds*

$$|X^\alpha(\rho) - X^\alpha(\widehat{\rho})| \leq CW_2(\rho, \widehat{\rho}), \quad (4.22)$$

where W_2 is the 2-Wasserstein distance, and C depends only on α, L, $\int_{\mathbb{R}^d} |x|^4 \rho(dx)$, and $\int_{\mathbb{R}^d} |x|^4 \widehat{\rho}(dx)$.

Finally let us prove Theorem 3:

Proof: (Theorem 3) By Theorem 4.3, each subsequence $\{\overline{X}^{m_k}\}_{k\in\mathbb{N}}$ with $m_0 \le 1/2$ and m_k converging to 0 as $k \to \infty$ admits a subsequence (denoted w.l.o.g. by itself) that converges weakly. By Skorokhod's lemma (see Ref. [85, Theorem 6.7, page 70]) and the existence and uniqueness of strong solution to SDE (4.1), we may find a common probability space $(\Omega, \mathcal{F}, \mathbb{P})$ on which the joint processes $\{(\overline{X}^{m_k}, B)\}_{k\in\mathbb{N}}$ converge to some process $(\widehat{X}, B)$ as random variables valued in $\mathcal{C}([0,T]; \mathbb{R}^{2d})$ almost surely. Here B is an identical d-dimensional Wiener process on $(\Omega, \mathcal{F}, \mathbb{P})$. In particular, we have

$$\mathbb{P}\left(\lim_{k\to\infty} \sup_{t\in[0,T]} |\overline{X}_t^{m_k} - \widehat{X}_t| = 0\right) = 1. \tag{4.23}$$

We shall verify that the limit $\widehat{X}$ is indeed the unique solution $\overline{X}$ to SDE (4.8).

Recalling the existence and uniqueness of the strong solution $\overline{X}^{m_k}$ to SDE (4.13) in Theorem 4.2, we have

$$\overline{X}_t^{m_k} = \overline{X}_0 + \frac{m_k}{\gamma}(1 - e^{-\frac{\gamma}{m_k}t})\overline{V}_0 + \frac{\lambda}{\gamma}\int_0^t (1 - e^{-\frac{\gamma}{m_k}(t-s)})(X^\alpha(\rho_s^{m_k}) - \overline{X}_s^{m_k})ds \tag{4.24}$$

$$+ \frac{\sigma}{\gamma}\int_0^t (1 - e^{-\frac{\gamma}{m_k}(t-s)})D(X^\alpha(\rho_s^{m_k}) - \overline{X}_s^{m_k})dB_s.$$

By the estimates in (4.18) and Fatou's lemma there exists a constant C_2 being independent of m_k such that

$$\sup_{k\in\mathbb{N}}\sup_{t\in[0,T]} \mathbb{E}\left[|\overline{X}_t^{m_k}|^4\right] + \sup_{t\in[0,T]} \mathbb{E}\left[|\widehat{X}_t|^4\right]$$
$$\le C_2 := C(\mathbb{E}[|\overline{X}_0|^4], \mathbb{E}[|\overline{V}_0|^4], C_{\alpha,\mathcal{F}}, \lambda, \sigma, d, T) < \infty. \tag{4.25}$$

As a straightforward consequence of the above boundedness, it holds that

$$\sup_{k\in\mathbb{N}, t\in[0,T]} \mathbb{P}(|\overline{X}_t^{m_k} - \widehat{X}_t| > A) \le \frac{2^4 C_2}{A^4}, \quad \forall A > 0. \tag{4.26}$$

Thus, the dominated convergence theorem gives that for each $A > 0$,

$$\lim_{k\to\infty} \mathbb{E}\left[\int_0^T |\overline{X}_t^{m_k} - \widehat{X}_t|^2\, dt\right]$$
$$\le \limsup_{k\to\infty}\left(\mathbb{E}\left[\int_0^T |\overline{X}_t^{m_k} - \widehat{X}_t|^2 \wedge A^2\, dt\right]\right)$$

$$+ \mathbb{E}\left[\int_0^T |\overline{X}_t^{m_k} - \widehat{X}_t|^2 1_{\{|\overline{X}_t^{m_k} - \widehat{X}_t| > A\}}\, dt\right]\Bigg)$$

$$\leq \limsup_{k \to \infty} \mathbb{E}\left[\int_0^T |\overline{X}_t^{m_k} - \widehat{X}_t|^2 \wedge A^2\, dt\right]$$

$$+ T \cdot \sup_{k \in \mathbb{N}} \sup_{t \in [0,T]} \left(\mathbb{E}\left[|\overline{X}_t^{m_k} - \widehat{X}_t|^4\right]\right)^{1/2} \left|\mathbb{P}(|\overline{X}_t^{m_k} - \widehat{X}_t| > A)\right|^{1/2}$$

$$\leq \limsup_{k \to \infty} \mathbb{E}\left[\int_0^T |\overline{X}_t^{m_k} - \widehat{X}_t|^2 \wedge A^2\, dt\right] + \frac{2^4 C_2 T}{A^2}$$

$$= \frac{2^4 C_2 T}{A^2},$$

which by the arbitrariness of $A > 0$ indicates that

$$\lim_{k \to \infty} \mathbb{E}\left[\int_0^T |\overline{X}_t^{m_k} - \widehat{X}_t|^2\, dt\right] = 0. \tag{4.27}$$

Letting $\rho(t, dx)$ be the probability distribution of $\widehat{X}_t$ for $t \in [0, T]$, Lemma 3.5 gives

$$|X^\alpha(\rho_t)| \leq (b_1 + b_2 \mathbb{E}[|\widehat{X}_t|^2])^{\frac{1}{2}} \leq (b_1 + b_2 C_2^{\frac{1}{2}})^{\frac{1}{2}} =: C_3,$$

and thus

$$\sup_{k \in \mathbb{N}} \sup_{t \in [0,T]} |X^\alpha(\rho_t^{m_k})| \leq C_3, \quad \text{and} \quad \sup_{t \in [0,T]} |X^\alpha(\rho_t)| \leq C_3. \tag{4.28}$$

Then we compare the SDEs (4.8) and (4.24) term by term. By Lemma 4.4, we have

$$|X^\alpha(\rho_t^{m_k}) - X^\alpha(\rho_t)|^2 \leq C W_2^2(\rho_t^{m_k}, \rho_t) \leq C \mathbb{E}[|\overline{X}_t^{m_k} - \widehat{X}_t|^2],$$

and thus by using the fact that $\gamma = 1 - m_k$, one has

$$\mathbb{E}\left[\left|\frac{\lambda}{\gamma}\int_0^t (1 - e^{-\frac{\gamma}{m_k}(t-s)})(X^\alpha(\rho_s^{m_k}) - \overline{X}_s^{m_k})ds - \lambda\int_0^t (X^\alpha(\rho_s) - \widehat{X}_s)ds\right|^2\right]$$

$$\leq 2\mathbb{E}\left[\left|\frac{\lambda}{1 - m_k}\int_0^t (1 - e^{-\frac{1 - m_k}{m_k}(t-s)}) \cdot (X^\alpha(\rho_s^{m_k}) - X^\alpha(\rho_s) + \widehat{X}_s - \overline{X}_s^{m_k})ds\right|^2\right]$$

$$+ 2\mathbb{E}\left[\left|\lambda\int_0^t \left(\frac{1 - e^{-\frac{1 - m_k}{m_k}(t-s)}}{1 - m_k} - 1\right)(X^\alpha(\rho_s) - \widehat{X}_s)ds\right|^2\right]$$

$$\leq C\mathbb{E}\left[\int_0^t |\widehat{X}_s - \overline{X}_s^{m_k}|^2\, ds\right] + C\lambda^2 \int_0^t \left|\frac{1 - e^{-\frac{1 - m_k}{m_k}(t-s)}}{1 - m_k} - 1\right|^2\, ds$$

$$\cdot \mathbb{E}\left[\int_0^T \left|X^\alpha(\rho_s) - \widehat{X}_s\right|^2 ds\right]$$

$$\leq C\mathbb{E}\left[\int_0^t \left|\widehat{X}_s - \overline{X}_s^{m_k}\right|^2 ds\right] + C\int_0^t \left|\frac{1 - e^{-\frac{1-m_k}{m_k}(t-s)} - (1 - m_k)}{1 - m_k}\right|^2 ds$$

$$\leq C\mathbb{E}\left[\int_0^t \left|\widehat{X}_s - \overline{X}_s^{m_k}\right|^2 ds\right] + C\int_0^t \left(|m_k|^2 + e^{-\frac{2(1-m_k)}{m_k}(t-s)}\right) ds$$

$$\leq C\mathbb{E}\left[\int_0^t \left|\widehat{X}_s - \overline{X}_s^{m_k}\right|^2 ds\right] + C\left(t\,|m_k|^2 + \frac{m_k}{2(1 - m_k)}\right), \tag{4.29}$$

where the constants Cs are independent of k. For the stochastic integrals, it holds analogously that

$$\mathbb{E}\left[\left|\frac{\sigma}{\gamma}\int_0^t (1 - e^{-\frac{\gamma}{m_k}(t-s)})D(X^\alpha(\rho_s^{m_k}) - \overline{X}_s^{m_k})dB_s\right.\right.$$

$$\left.\left. - \sigma\int_0^t D(X^\alpha(\rho_s) - \widehat{X}_s)dB_s\right|^2\right]$$

$$\leq d\sigma^2 \sum_{n=1}^d \mathbb{E}\left[\left|\frac{1}{\gamma}\int_0^t (1 - e^{-\frac{\gamma}{m_k}(t-s)})(X^\alpha(\rho_s^{m_k}) - \overline{X}_s^{m_k})_n dB_s^n e_n\right.\right.$$

$$\left.\left. - \int_0^t (X^\alpha(\rho_s) - \widehat{X}_s)_n dB_s^n e_n\right|^2\right]$$

$$= d\sigma^2 \sum_{n=1}^d \mathbb{E}\left[\int_0^t \left|\frac{1 - e^{-\frac{\gamma}{m_k}(t-s)}}{\gamma}(X^\alpha(\rho_s^{m_k}) - \overline{X}_s^{m_k})_n\right.\right.$$

$$\left.\left. - (X^\alpha(\rho_s) - \widehat{X}_s)_n\right|^2 ds\right]. \tag{4.30}$$

Thus we have

$$\mathbb{E}\left[\left|\frac{\sigma}{\gamma}\int_0^t (1 - e^{-\frac{\gamma}{m_k}(t-s)})D(X^\alpha(\rho_s^{m_k}) - \overline{X}_s^{m_k})dB_s\right.\right.$$

$$\left.\left. - \sigma\int_0^t D(X^\alpha(\rho_s) - \widehat{X}_s)dB_s\right|^2\right]$$

$$\leq 2d\sigma^2 \sum_{n=1}^d \mathbb{E}\left[\int_0^t \left|\frac{1 - e^{-\frac{\gamma}{m_k}(t-s)}}{\gamma}\right.\right.$$

$$\left.\left. \left((X^\alpha(\rho_s^{m_k}) - \overline{X}_s^{m_k})_n - (X^\alpha(\rho_s) - \widehat{X}_s)_n\right)\right|^2 ds\right]$$

$$+ 2d\sigma^2 \sum_{n=1}^d \mathbb{E}\left[\int_0^t \left|\left(\frac{1 - e^{-\frac{\gamma}{m_k}(t-s)}}{\gamma} - 1\right)(X^\alpha(\rho_s) - \widehat{X}_s)_n\right|^2 ds\right]$$

$$\leq C\mathbb{E}\left[\int_0^t \left|\overline{X}_s^{m_k} - \widehat{X}_s\right|^2 ds\right]$$

$$+ 2d\sigma^2 \sup_{s\in[0,t]} \mathbb{E}\left[\left|(X^\alpha(\rho_s) - \widehat{X}_s)\right|^2\right] \cdot \int_0^t \left|\left(\frac{1 - e^{-\frac{\gamma}{m_k}(t-s)}}{\gamma} - 1\right)\right|^2 ds$$

$$\leq C\mathbb{E}\left[\int_0^t \left|\widehat{X}_s - \overline{X}_s^{m_k}\right|^2 ds\right] + C\left(t\,|m_k|^2 + \frac{m_k}{2(1-m_k)}\right). \tag{4.31}$$

In addition, it is obvious that

$$\left|\frac{m_k}{\gamma}(1 - e^{-\frac{\gamma}{m_k}t})\overline{V}_0\right| \leq Cm_k\,|\overline{V}_0|. \tag{4.32}$$

Combining the estimates (4.29)–(4.32), letting k tend to infinity on both sides of (4.24) and recalling $\frac{1}{2} \geq m_k \to 0^+$ and the relation (4.27), we have

$$\widehat{X}_t = \overline{X}_0 + \lambda \int_0^t (X^\alpha(\rho_s) - \widehat{X}_s)ds + \sigma \int_0^t D(X^\alpha(\rho_s) - \widehat{X}_s)dB_s.$$

Therefore, the limit $\widehat{X}$ turns out to be a solution to SDE (4.8). Meanwhile, in view of the continuity of $X^\alpha(\rho)$ in Lemma 4.4, we can easily show that (4.8) admits a unique (strong) solution as in Theorem 4.2 by using Leray-Schauder fixed point theorem as in Ref. [51, Theorem 3.1]. Thus, we must have $\widehat{X} = \overline{X}$ that is the unique strong solution to SDE (4.8) with $\sup_{t\in[0,T]} \mathbb{E}\left[|\overline{X}_t|^4\right] \leq C_2$. Further, due to the arbitrariness of the subsequence $\{\overline{X}^{m_k}\}_{k\in\mathbb{N}}$, we conclude that as $m \to 0^+$, the sequence of stochastic processes $\{\overline{X}^m\}_{0<m\leq\frac{1}{2}}$ converge weakly to the unique solution $\overline{X}$ to SDE (4.8).

Finally, to measure the distance between $\overline{X}^m$ and the limit $\widehat{X} = \overline{X}$, we may have similar calculations to (4.29)–(4.32), subtract both sides of SDEs (4.8) from those of (4.24), and arrive at

$$\mathbb{E}[|\overline{X}_t^m - \overline{X}_t|^2] \leq C\int_0^t \mathbb{E}[|\overline{X}_s^m - \overline{X}_s|^2]ds + C\,m, \quad t \in [0,T].$$

By Gronwall's inequality it implies that

$$\sup_{t\in[0,T]} \mathbb{E}[|\overline{X}_t^m - \overline{X}_t|^2] \leq Cm, \tag{4.33}$$

where C depends only on $\mathbb{E}[|\overline{X}_0|^4], \mathbb{E}[|\overline{V}_0|^4], C_u, M, C_l, \lambda, \sigma, d$, and T. This completes the proof. $\qquad\square$

4.2. The general case with memory

Next, we consider the same small inertia scaling in the general case with dependence from the local best. Again, we first write down the nonlinear McKean-Vlasov process corresponding to the SD-PSO system (3.33), which is of the form

$$d\overline{X}_t^m = \overline{V}_t^m dt, \tag{4.34a}$$

$$d\overline{Y}_t^m = \nu\left(\overline{X}_t^m - \overline{Y}_t^m\right) S^\beta\left(\overline{X}_t^m, \overline{Y}_t^m\right) dt, \tag{4.34b}$$

$$d\overline{V}_t^m = -\frac{\gamma}{m}\overline{V}_t^m dt + \frac{\lambda_1}{m}\left(\overline{Y}_t^m - \overline{X}_t^m\right) dt$$

$$\qquad + \frac{\lambda_2}{m}\left(Y^\alpha(\overline{\rho}_t^m) - \overline{X}_t^m\right) dt + \frac{\sigma_1}{m}D\left(\overline{Y}_t^m - \overline{X}_t^m\right) dB_t^1 \tag{4.34c}$$

$$\qquad + \frac{\sigma_2}{m}D\left(Y^\alpha(\overline{\rho}_t^m) - \overline{X}_t^m\right) dB_t^2,$$

where B^1 and B^2 are two mutually independent d-dimensional Wiener processes, and similarly to the previous section, we introduce the following regularization of the global best position

$$Y^\alpha(\overline{\rho}_t^m) = \frac{\int_{\mathbb{R}^d} y\omega_\alpha(y)\overline{\rho}^m(t, dy)}{\int_{\mathbb{R}^d} \omega_\alpha(y)\overline{\rho}^m(t, dy)}, \qquad \overline{\rho}^m(t, y) = \iint_{\mathbb{R}^d \times \mathbb{R}^d} f^m(t, dx, y, dv).$$

As $m \to 0^+$ we formally get from (4.34c)

$$\overline{V}_t^0 dt = \lambda_1\left(\overline{Y}_t^0 - \overline{X}_t^0\right) dt + \lambda_2\left(Y^\alpha(\overline{\rho}_t^0) - \overline{X}_t^0\right) dt$$

$$\qquad + \sigma_1 D(\overline{Y}_t^0 - \overline{X}_t^0)dB_t^1 + \sigma_2 D(Y^\alpha(\overline{\rho}_t^0) - \overline{X}_t^0)dB_t^2,$$

which inserted into (4.34a) and omitting the superscripts corresponds to a novel *CBO system with local best*

$$d\overline{X}_t = \lambda_1(\overline{Y}_t - \overline{X}_t)dt + \lambda_2(Y^\alpha(\overline{\rho}_t) - \overline{X}_t)dt$$

$$\qquad + \sigma_1 D(\overline{Y}_t - \overline{X}_t)dB_t^1 + \sigma_2 D(Y^\alpha(\overline{\rho}_t) - \overline{X}_t)dB_t^2, \tag{4.35}$$

$$d\overline{Y}_t = \nu\left(\overline{X}_t - \overline{Y}_t\right) S^\beta\left(\overline{X}_t, \overline{Y}_t\right) dt.$$

In contrast with the model recently introduced in Ref. [53] the above first order CBO method avoids backward time integration through the use of an additional differential equation. We refer to Ref. [67] for further details on the above CBO system.

4.2.1. Formal derivation in the mean-field case

Concerning the corresponding MF-PSO limit characterized by (3.36) for $m \to 0^+$ we can essentially perform analogous computations as in the previous section (see Ref. [68]). Similarly by considering the local Maxwellian with unitary mass and zero momentum

$$\mathcal{M}_m(x, y, v, t) = \prod_{j=1}^{d} M_m(x_j, y_j, v_j, t),$$

$$M_m(x_j, y_j, v_j, t) = \frac{m^{1/2}}{\pi^{1/2} |\Sigma(x_j, y_j, t)|} \exp\left\{ -\frac{mv_j^2}{\Sigma(x_j, y_j, t)^2} \right\},$$

where

$$\Sigma(x_j, y_j, t)^2 = \sigma_2^2 (x_j - Y_j^\alpha(\overline{\rho}))^2 + \sigma_1^2 (x_j - y_j)^2,$$

we can assume for $m \ll 1$

$$f(x, y, v, t) = \rho(x, y, t) \mathcal{M}_\varepsilon(x, y, v, t). \tag{4.36}$$

After integration of the MF-PSO equation (3.36) with respect to v, we get the second order *macroscopic PSO system with local best*

$$\frac{\partial \rho}{\partial t} + \nabla_x \cdot (\rho u) + \nabla_y \cdot \left(\nu(x - y) S^\beta(x, y) \rho \right) = 0$$

$$\frac{\partial (\rho u)_j}{\partial t} + \frac{\sigma^2}{2m} \frac{\partial}{\partial x_j} \left(\rho(x, t) \Sigma(x_j, y_j, t)^2 \right) = \tag{4.37}$$

$$-\frac{\gamma}{m} (\rho u)_j + \frac{1}{m} (\lambda_1 (y_j - x_j) + \lambda_2 (Y_j^\alpha(\overline{\rho}) - x_j)) \rho.$$

Formally, as $m \to 0^+$, the above system reduces to a novel *mean-field CBO system with local best*

$$\frac{\partial \rho}{\partial t} + \nabla_x \cdot (\lambda_1 (y - x) + \lambda_2 (Y^\alpha(\overline{\rho}) - x)) \rho$$

$$+ \nabla_y \cdot \left(\nu(x - y) S^\beta(x, y) \rho \right)$$

$$= \frac{1}{2} \sum_{j=1}^{d} \frac{\partial^2}{\partial x_j^2} \left(\rho(x, t) \left(\sigma_1^2 (x_j - y_j)^2 + \sigma_2^2 (x_j - Y_j^\alpha(\overline{\rho}))^2 \right) \right). \tag{4.38}$$

4.2.2. *Rigorous derivation*

Since the proof of the zero-inertia limit for the PSO dynamics with memory effects follows similar arguments as developed in Sec. 4.1.2 and no essential innovation is needed to be explained, we only recall the main results here.

Let us solve (4.34c) to obtain

$$
\begin{aligned}
\overline{X}_t^m = \overline{X}_0 &+ \frac{m}{\gamma}(1 - e^{-\frac{\gamma}{m}t})\overline{V}_0 + \frac{\lambda_1}{\gamma}\int_0^t (1 - e^{-\frac{\gamma}{m}(t-s)})(\overline{Y}_s^m - \overline{X}_s^m)ds \\
&+ \frac{\sigma_1}{\gamma}\int_0^t (1 - e^{-\frac{\gamma}{m}(t-s)})D(\overline{Y}_s^m - \overline{X}_s^m)dB_s^1 \\
&+ \frac{\lambda_2}{\gamma}\int_0^t (1 - e^{-\frac{\gamma}{m}(t-s)})(Y^\alpha(\overline{\rho}_s^m) - \overline{X}_s^m)ds \\
&+ \frac{\sigma_2}{\gamma}\int_0^t (1 - e^{-\frac{\gamma}{m}(t-s)})D(Y^\alpha(\overline{\rho}_s^m) - \overline{X}_s^m)dB_s^2
\end{aligned}
\tag{4.39}
$$

and

$$
\overline{Y}_t^m = \overline{Y}_0 + \nu \int_0^t \left(\overline{X}_s^m - \overline{Y}_s^m\right) S^\beta \left(\overline{X}_s^m, \overline{Y}_s^m\right) ds.
\tag{4.40}
$$

Similar to Theorem 4.3 one can prove the following result of tightness.

Theorem 4.5: *Let Assumption 1 hold and* $(\overline{X}_t^m, \overline{Y}_t^m, \overline{V}_t^m)_{t\in[0,T]}$ *satisfy the system* (4.34a)–(4.34c). *For each countable subsequence* $\{m_k\}_{k\in\mathbb{N}} \subset [0, \frac{1}{2}]$ *with* $\lim_{k\to\infty} m_k = 0$, *the sequence of probability distributions* $\{\rho^{m_k}\}_{k\in\mathbb{N}}$ *of* $\{(\overline{X}^{m_k}, \overline{Y}^{m_k})\}_{k\in\mathbb{N}}$ *is tight.*

Then following the lines of the proof in Theorem 3, one can obtain

---

Theorem 4: *Let Assumption 1 hold and* $(\overline{X}_t^m, \overline{Y}_t^m)_{t\in[0,T]}$ *satisfy the system* (4.39)–(4.40). *Then as* $m \to 0^+$, *the sequence of stochastic processes* $\{(\overline{X}^m, \overline{Y}^m)\}_{0<m\leq\frac{1}{2}}$ *converge weakly to* $(\overline{X}, \overline{Y})$ *which is the unique solution to the following coupled SDE:*

$$
\begin{aligned}
\overline{X}_t = \overline{X}_0 &+ \lambda_1 \int_0^t (\overline{Y}_s - \overline{X}_s)ds + \sigma_1 \int_0^t D(\overline{Y}_s - \overline{X}_s)dB_s^1 \\
&+ \lambda_2 \int_0^t (Y^\alpha(\overline{\rho}_s) - \overline{X}_s)ds + \sigma_2 \int_0^t D(Y^\alpha(\overline{\rho}_s) - \overline{X}_s)dB_s^2, \\
\overline{Y}_t = \overline{Y}_0 &+ \nu \int_0^t (\overline{X}_s - \overline{Y}_s) S^\beta (\overline{X}_s, \overline{Y}_s) ds.
\end{aligned}
$$

Moreover it holds that

$$\sup_{t\in[0,T]} \mathbb{E}\left[|\overline{X}_t^m - \overline{X}_t|^2 + |\overline{Y}_t^m - \overline{Y}_t|^2\right] \leq C\,m, \tag{4.41}$$

where the constant C depends only on $\mathbb{E}[|\overline{X}_0|^4]$, $\mathbb{E}[|\overline{Y}_0|^4]$, $\mathbb{E}[|\overline{V}_0|^4]$, λ_1, σ_2, λ_2, σ_2, d, β, T, C_u, M, C_l, and ν.

5. Convergence to the global minimum

In this section we present some results on the global convergence of the PSO model (3.1) without memory effects. The extension to the case with memory effects is not straightforward and is actually under study. Here we will follow the presentation in Ref. [72], we refer to Refs. [51,52,61–63] for similar results for CBO and related models. A different approach to the global convergence of CBO has been presented recently in Ref. [58].

Let $(\overline{X}_t, \overline{V}_t)_{t\geq 0}$ be the solution to the nonlinear SDE (4.1) (dropping the superscript m), and consider the quantity

$$\mathcal{H}(t) := \left(\frac{\gamma}{2m}\right)^2 |\overline{X}_t - \mathbb{E}[\overline{X}_t]|^2 + |\overline{V}_t|^2 + \frac{\gamma}{2m}(\overline{X}_t - \mathbb{E}[\overline{X}_t]) \cdot \overline{V}_t,$$

then it holds that

$$\mathcal{H}(t) \geq \frac{1}{2}\left(\frac{\gamma}{2m}\right)^2 |\overline{X}_t - \mathbb{E}[\overline{X}_t]|^2 + \frac{1}{2}|\overline{V}_t|^2$$

$$\mathcal{H}(t) \leq \frac{3}{2}\left(\frac{\gamma}{2m}\right)^2 |\overline{X}_t - \mathbb{E}[\overline{X}_t]|^2 + \frac{3}{2}|\overline{V}_t|^2 \tag{5.1}$$

$$\leq \frac{3}{2}\left(\left(\frac{\gamma}{2m}\right)^2 + 1\right)\left(|\overline{X}_t - \mathbb{E}[\overline{X}_t]|^2 + |\overline{V}_t|^2\right).$$

The goal is then to obtain the decay property of $\mathcal{H}(t)$.

In the following we shall use the notation

$$\delta\overline{X}_t := \overline{X}_t - \mathbb{E}[\overline{X}_t], \tag{5.2}$$

then $\mathbb{E}[|\delta\overline{X}_t|^2]$ is the variance of X_t. Now we can derive an evolution inequality of the quantity $\mathbb{E}[\mathcal{H}(t)]$.

Theorem 5.1: *Under the Assumption 1, let $(\overline{X}_t, \overline{V}_t)_{t\geq 0}$ be the solution to the nonlinear SDE (4.1). Then $\mathbb{E}[\mathcal{H}(t)]$ satisfies*

$$\frac{d}{dt}\mathbb{E}[\mathcal{H}(t)] \leq -\frac{\gamma}{m}\mathbb{E}[|\overline{V}_t|^2]$$

$$-\left(\frac{\lambda\gamma}{2m^2} - \left(\frac{2\lambda^2}{\gamma m} + \frac{\sigma^2}{m^2}\right)\frac{2e^{-\alpha\mathcal{F}}}{\mathbb{E}[e^{-\alpha\mathcal{F}(\overline{X}_t)}]}\right)\mathbb{E}[|\delta\overline{X}_t|^2]. \tag{5.3}$$

Proof: First the integration by parts formula gives

$$\frac{d}{dt}\mathbb{E}[|\delta\overline{X}_t|^2] = 2\mathbb{E}[\delta\overline{X}_t \cdot \overline{V}_t], \tag{5.4}$$

where we have used the fact that $\mathbb{E}[\delta\overline{X}_t \cdot \mathbb{E}[V_t]] = 0$. Applying Itô-Doeblin formula and taking zero-value of the stochastic integrals, we have for any $\varepsilon > 0$,

$$\frac{d}{dt}\mathbb{E}[|\overline{V}_t|^2] = -2\frac{\gamma}{m}\mathbb{E}[|\overline{V}_t|^2] + 2\frac{\lambda}{m}\mathbb{E}[\overline{V}_t \cdot (X^\alpha(\rho_t) - \overline{X}_t)]$$

$$+ \frac{\sigma^2}{m^2}\mathbb{E}[|X^\alpha(\rho_t) - \overline{X}_t|^2]$$

$$\leq -\left(\frac{2\gamma}{m} - \frac{\lambda}{\varepsilon m}\right)\mathbb{E}[|\overline{V}_t|^2] + \left(\frac{\varepsilon\lambda}{m} + \frac{\sigma^2}{m^2}\right)\mathbb{E}[|X^\alpha(\rho_t) - \overline{X}_t|^2]. \tag{5.5}$$

Further by Itô-Doeblin formula, it holds that

$$\frac{d}{dt}\mathbb{E}[\delta\overline{X}_t \cdot \overline{V}_t]$$

$$= \mathbb{E}[|\overline{V}_t|^2] - (\mathbb{E}[\overline{V}_t])^2 - \frac{\gamma}{m}\mathbb{E}[\delta\overline{X}_t \cdot \overline{V}_t] + \frac{\lambda}{m}\mathbb{E}[\delta\overline{X}_t \cdot (X^\alpha(\rho_t) - \overline{X}_t)]$$

$$\leq \mathbb{E}[|\overline{V}_t|^2] - \frac{\gamma}{2m}\frac{d}{dt}\mathbb{E}[|\delta\overline{X}_t|^2] - \frac{\lambda}{m}\mathbb{E}[|\delta\overline{X}_t|^2] + \frac{\lambda}{m}\mathbb{E}[\delta\overline{X}_t \cdot (X^\alpha(\rho_t) - \mathbb{E}[\overline{X}_t])]$$

$$= \mathbb{E}[|\overline{V}_t|^2] - \frac{\gamma}{2m}\frac{d}{dt}\mathbb{E}[|\delta\overline{X}_t|^2] - \frac{\lambda}{m}\mathbb{E}[|\delta\overline{X}_t|^2], \tag{5.6}$$

where we have used (5.4) and the fact that $\mathbb{E}[\delta\overline{X}_t \cdot (X^\alpha(\rho_t) - \mathbb{E}[\overline{X}_t])] = 0$. Thus, we have

$$\left(\frac{\gamma}{2m}\right)^2 \frac{d}{dt}\mathbb{E}[|\delta\overline{X}_t|^2] + \frac{\gamma}{2m}\frac{d}{dt}\mathbb{E}[\delta\overline{X}_t \cdot \overline{V}_t]$$

$$\leq \frac{\gamma}{2m}\mathbb{E}[|\overline{V}_t|^2] - \frac{\lambda\gamma}{2m^2}\mathbb{E}[|\delta\overline{X}_t|^2]. \tag{5.7}$$

Collecting estimates (5.5) and (5.7) yields that

$$\frac{d}{dt}\mathbb{E}[\mathcal{H}(t)] \leq -\left(\frac{2\gamma}{m} - \frac{\lambda}{\varepsilon m} - \frac{\gamma}{2m}\right)\mathbb{E}[|\overline{V}_t|^2] - \frac{\lambda\gamma}{2m^2}\mathbb{E}[|\delta\overline{X}_t|^2]$$

$$+ \left(\frac{\varepsilon\lambda}{m} + \frac{\sigma^2}{m^2}\right)\mathbb{E}[|X^\alpha(\rho_t) - \overline{X}_t|^2]. \tag{5.8}$$

To estimate the term $\mathbb{E}[|\overline{X}_t - X^\alpha(\rho_t)|^2]$, we apply Jensen's inequality to obtain

$$\mathbb{E}[|\overline{X}_t - X^\alpha(\rho_t)|^2] \leq \frac{\iint |x - y|^2 \omega_\alpha^{\mathcal{F}}(y)\rho_t(dy)\rho_t(dx)}{\int \omega_\alpha^{\mathcal{F}}(y)\rho_t(dy)}$$

$$\leq 2e^{-\alpha\underline{\mathcal{F}}} \frac{\mathbb{E}[|\delta\overline{X}_t|^2]}{\mathbb{E}[e^{-\alpha\mathcal{F}(\overline{X}_t)}]} . \qquad (5.9)$$

Hence, by choosing $\varepsilon = \frac{2\lambda}{\gamma}$ we obtain

$$\frac{d}{dt}\mathbb{E}[\mathcal{H}(t)] \leq -\frac{\gamma}{m}\mathbb{E}[|\overline{V}_t|^2]$$

$$-\left(\frac{\lambda\gamma}{2m^2} - \left(\frac{2\lambda^2}{\gamma m} + \frac{\sigma^2}{m^2}\right)\frac{2e^{-\alpha\underline{\mathcal{F}}}}{\mathbb{E}[e^{-\alpha\mathcal{F}(\overline{X}_t)}]}\right)\mathbb{E}[|\delta\overline{X}_t|^2], \quad (5.10)$$

which completes the proof. $\square$

Next we study the evolution of the quantity $\mathbb{E}[e^{-\alpha\mathcal{F}(\overline{X}_t)}]$, and we need an additional assumption on the cost function $\mathcal{F}$ that

A1: $\mathcal{F} \in C^2(\mathbb{R}^d)$ with $\|\nabla^2\mathcal{F}\|_\infty \leq c_{\mathcal{F}}$ for some constant $c_{\mathcal{F}} > 0$.

Lemma 5.2: *Under the Assumption 1 and **A1**, let $(\overline{X}_t, \overline{V}_t)_{t\geq 0}$ be the solution to the nonlinear SDE (4.1). Then it holds that*

$$\frac{d^2}{dt^2}(\mathbb{E}[e^{-\alpha\mathcal{F}(\overline{X}_t)}])^2 \geq -\frac{\gamma}{m}\frac{d}{dt}(\mathbb{E}[e^{-\alpha\mathcal{F}(\overline{X}_t)}])^2$$

$$-4\left(\alpha + \frac{\alpha\lambda}{m}2\left(\frac{2m}{\gamma}\right)^2\right)c_{\mathcal{F}}e^{-2\alpha\underline{\mathcal{F}}}\mathbb{E}[\mathcal{H}(t)]. \quad (5.11)$$

Proof: First, applying Itô-Doeblin formula and taking zero-value of the stochastic integrals, we have

$$\frac{d}{dt}\mathbb{E}[e^{-\alpha\mathcal{F}(\overline{X}_t)}] = -\alpha\mathbb{E}[e^{-\alpha\mathcal{F}(\overline{X}_t)}\nabla\mathcal{F}(\overline{X}_t) \cdot \overline{V}_t]$$

$$= -\alpha\mathbb{E}\left[\int_0^t d\langle e^{-\alpha\mathcal{F}(\overline{X}_s)}\nabla\mathcal{F}(\overline{X}_s), \overline{V}_s\rangle\right] + \alpha\mathbb{E}[e^{-\alpha\mathcal{F}(\overline{X}_0)}\langle\nabla\mathcal{F}(\overline{X}_0), \overline{V}_0\rangle]$$

$$= \alpha\mathbb{E}[e^{-\alpha\mathcal{F}(\overline{X}_0)}\langle\nabla\mathcal{F}(\overline{X}_0), \overline{V}_0\rangle] - \alpha\mathbb{E}\left[\int_0^t \langle e^{-\alpha\mathcal{F}(\overline{X}_s)}\overline{V}_s\nabla^2\mathcal{F}(\overline{X}_s), \overline{V}_s\rangle ds\right]$$

$$+ \alpha^2\mathbb{E}\left[\int_0^t e^{-\alpha\mathcal{F}(\overline{X}_s)}|\langle\nabla\mathcal{F}(\overline{X}_s), \overline{V}_s\rangle|^2 ds\right]$$

$$- \alpha\mathbb{E}\left[\int_0^t e^{-\alpha\mathcal{F}(\overline{X}_s)}\langle\nabla\mathcal{F}(\overline{X}_s), -\frac{\gamma}{m}\overline{V}_s\rangle ds\right]$$

$$- \alpha\mathbb{E}\left[\int_0^t e^{-\alpha\mathcal{F}(\overline{X}_s)}\langle\nabla\mathcal{F}(\overline{X}_s), \frac{\lambda}{m}(X^\alpha(\rho_s) - \overline{X}_s)\rangle ds\right] .$$

Further, differentiating both sides with respect to t gives

$$\frac{d^2}{dt^2}\mathbb{E}[e^{-\alpha\mathcal{F}(\overline{X}_t)}] = -\alpha\mathbb{E}[\langle e^{-\alpha\mathcal{F}(\overline{X}_t)}\overline{V}_t\nabla^2\mathcal{F}(\overline{X}_t), \overline{V}_t\rangle]$$

$$+ \alpha^2\mathbb{E}[e^{-\alpha\mathcal{F}(\overline{X}_t)}|\langle\nabla\mathcal{F}(\overline{X}_t), \overline{V}_t\rangle|^2]$$

$$- \alpha\mathbb{E}\left[e^{-\alpha\mathcal{F}(\overline{X}_t)}\left\langle\nabla\mathcal{F}(\overline{X}_t), -\frac{\gamma}{m}\overline{V}_t\right\rangle\right]$$

$$- \alpha\mathbb{E}\left[e^{-\alpha\mathcal{F}(\overline{X}_t)}\left\langle\nabla\mathcal{F}(\overline{X}_t), \frac{\lambda}{m}(X^\alpha(\rho_t) - \overline{X}_t)\right\rangle\right]$$

$$\geq -\frac{\gamma}{m}\frac{d}{dt}\mathbb{E}[e^{-\alpha\mathcal{F}(\overline{X}_t)}] - \alpha\mathbb{E}[\langle e^{-\alpha\mathcal{F}(\overline{X}_t)}\overline{V}_t\nabla^2\mathcal{F}(\overline{X}_t), \overline{V}_t\rangle]$$

$$- \alpha\mathbb{E}\left[e^{-\alpha\mathcal{F}(\overline{X}_t)}\left\langle\nabla\mathcal{F}(\overline{X}_t), \frac{\lambda}{m}(X^\alpha(\rho_t) - \overline{X}_t)\right\rangle\right]$$

$$=: -\frac{\gamma}{m}\frac{d}{dt}\mathbb{E}[e^{-\alpha\mathcal{F}(\overline{X}_t)}] + I_1 + I_2\,, \tag{5.12}$$

where one has used the fact that

$$-\frac{\gamma}{m}\frac{d}{dt}\mathbb{E}[e^{-\alpha\mathcal{F}(\overline{X}_t)}] = \frac{\alpha\gamma}{m}\mathbb{E}[e^{-\alpha\mathcal{F}(\overline{X}_t)}\langle\nabla\mathcal{F}(\overline{X}_t), \overline{V}_t\rangle]\,. \tag{5.13}$$

According to assumption **A1**, it is easy to see that

$$I_1 \geq -\alpha\mathbb{E}[e^{-\alpha\mathcal{F}(\overline{X}_t)}\|\nabla^2\mathcal{F}(\overline{X}_t)\|_\infty|\overline{V}_t|^2] \geq -\alpha c_\mathcal{F}e^{-\alpha\underline{\mathcal{F}}}\mathbb{E}[|\overline{V}_t|^2]. \tag{5.14}$$

We further notice that

$$\left|\mathbb{E}[e^{-\alpha\mathcal{F}(\overline{X}_t)}\langle\nabla\mathcal{F}(\overline{X}_t), (X^\alpha(\rho_t) - \overline{X}_t)\rangle]\right|$$

$$= \left|\mathbb{E}[e^{-\alpha\mathcal{F}(\overline{X}_t)}\langle\nabla\mathcal{F}(\overline{X}_t) - \nabla\mathcal{F}(X^\alpha(\rho_t)), (\overline{X}_t - X^\alpha(\rho_t))\rangle]\right|$$

$$\leq e^{-\alpha\underline{\mathcal{F}}}c_\mathcal{F}\mathbb{E}[|\overline{X}_t - X^\alpha(\rho_t)|^2]\,, \tag{5.15}$$

where we have used the fact that $\mathbb{E}[e^{-\alpha\mathcal{F}(\overline{X}_t)}\langle\nabla\mathcal{F}(X^\alpha(\rho_t)), (\overline{X}_t - X^\alpha(\rho_t))\rangle] = 0$. Furthermore since $\mathbb{E}[|\overline{X}_t - X^\alpha(\rho_t)|^2] \leq 2e^{-\alpha\underline{\mathcal{F}}}\frac{\mathbb{E}[|\delta\overline{X}_t|^2]}{\mathbb{E}[e^{-\alpha\mathcal{F}(\overline{X}_t)}]}$, one has

$$I_2 \geq -\frac{\alpha\lambda}{m}e^{-\alpha\underline{\mathcal{F}}}c_\mathcal{F}\mathbb{E}[|\overline{X}_t - X^\alpha(\rho_t)|^2]$$

$$\geq -\frac{\alpha\lambda}{m}2e^{-2\alpha\underline{\mathcal{F}}}c_\mathcal{F}\frac{\mathbb{E}[|\delta\overline{X}_t|^2]}{\mathbb{E}[e^{-\alpha\mathcal{F}(\overline{X}_t)}]}\,. \tag{5.16}$$

This combining with (5.14) leads to

$$\frac{d^2}{dt^2}\mathbb{E}[e^{-\alpha\mathcal{F}(\overline{X}_t)}] \geq -\frac{\gamma}{m}\frac{d}{dt}\mathbb{E}[e^{-\alpha\mathcal{F}(\overline{X}_t)}] - \alpha c_\mathcal{F}e^{-\alpha\underline{\mathcal{F}}}\mathbb{E}[|\overline{V}_t|^2]$$

$$-\frac{\alpha\lambda}{m}c_{\mathcal{F}}2e^{-2\alpha\underline{\mathcal{F}}}\frac{\mathbb{E}[|\delta\overline{X}_t|^2]}{\mathbb{E}[e^{-\alpha\mathcal{F}(\overline{X}_t)}]} \, . \tag{5.17}$$

Using this, one can obtain

$$\frac{d}{dt}\left(\frac{1}{2}\frac{d}{dt}(\mathbb{E}[e^{-\alpha\mathcal{F}(\overline{X}_t)}])^2\right) = \frac{d}{dt}\left(\mathbb{E}[e^{-\alpha\mathcal{F}(\overline{X}_t)}]\frac{d}{dt}(\mathbb{E}[e^{-\alpha\mathcal{F}(\overline{X}_t)}])\right)$$

$$= \left(\frac{d}{dt}(\mathbb{E}[e^{-\alpha\mathcal{F}(\overline{X}_t)}])\right)^2 + \mathbb{E}[e^{-\alpha\mathcal{F}(\overline{X}_t)}]\frac{d^2}{dt^2}\mathbb{E}[e^{-\alpha\mathcal{F}(\overline{X}_t)}]$$

$$\geq -\frac{\gamma}{2m}\frac{d}{dt}(\mathbb{E}[e^{-\alpha\mathcal{F}(\overline{X}_t)}])^2 - \alpha c_{\mathcal{F}}e^{-2\alpha\underline{\mathcal{F}}}\mathbb{E}[|\overline{V}_t|^2]$$

$$-\frac{\alpha\lambda}{m}2e^{-2\alpha\underline{\mathcal{F}}}c_{\mathcal{F}}\mathbb{E}[|\delta\overline{X}_t|^2]$$

$$\geq -\frac{\gamma}{2m}\frac{d}{dt}(\mathbb{E}[e^{-\alpha\mathcal{F}(\overline{X}_t)}])^2$$

$$-\left(2\alpha + 2\frac{\alpha\lambda}{m}2\left(\frac{2m}{\gamma}\right)^2\right)c_{\mathcal{F}}e^{-2\alpha\underline{\mathcal{F}}}\mathbb{E}[\mathcal{H}(t)] \, , \tag{5.18}$$

where we have used (5.1) in the last inequality. This completes the proof.
□

Our main theorem on global convergence can be described in the following way:

Theorem 5: *Under the Assumption 1 and **A1**, let $(\overline{X}_t, \overline{V}_t)_{t\geq 0}$ be the solution to the nonlinear SDE (4.1). Further we assume that the initial data $\overline{X}_0$ and $\overline{V}_0$ satisfy*

$$\mu := \frac{\lambda\gamma}{2m^2} - \left(\frac{2\lambda^2}{\gamma m} + \frac{\sigma^2}{m^2}\right)\frac{4e^{-\alpha\underline{\mathcal{F}}}}{\mathbb{E}[e^{-\alpha\mathcal{F}(\overline{X}_0)}]} > 0, \tag{5.19}$$

and

$$2\alpha\frac{\gamma}{m}\mathbb{E}[e^{-\alpha\mathcal{F}(\overline{X}_0)}]\left(\mathbb{E}[e^{-\alpha\mathcal{F}(\overline{X}_0)}\nabla\mathcal{F}(\overline{X}_0)\cdot\overline{V}_0]\right)_+$$

$$+ 4\left(\alpha + \frac{\alpha\lambda}{m}2\left(\frac{2m}{\gamma}\right)^2\right)c_{\mathcal{F}}e^{-2\alpha\underline{\mathcal{F}}}\frac{\mathbb{E}[\mathcal{H}(0)]}{\chi(\frac{\gamma}{m} - \chi)} < \frac{3}{4}(\mathbb{E}[e^{-\alpha\mathcal{F}(\overline{X}_0)}])^2 \, , \tag{5.20}$$

where we denote $x_+ = \max\{x, 0\}$, $\forall x \in \mathbb{R}$, and

$$\chi = \frac{\min\{\mu, \frac{\gamma}{m}\}}{\frac{3}{2}((\frac{\gamma}{2m})^2 + 1)}.$$

Then $\mathbb{E}[|\overline{X}_t - \mathbb{E}[\overline{X}_t]|^2] \to 0, \mathbb{E}[|\overline{V}_t|^2] \to 0$ *exponentially fast as* $t \to \infty$, *and there exists some* $\widetilde{x}$ *depending on* α *such that* $\mathbb{E}[\overline{X}_t] \to \widetilde{x}$ *and* $X^\alpha(\rho_t) \to \widetilde{x}$ *exponentially fast as* $t \to \infty$. *Moreover it holds that*

$$\mathcal{F}(\widetilde{x}) - \underline{\mathcal{F}} \le \frac{1}{\alpha}\log(2) - \frac{1}{\alpha}\log(\mathbb{E}[e^{-\alpha\mathcal{F}(\overline{X}_0)}]) - \underline{\mathcal{F}} \to 0 \text{ as } \alpha \to \infty. \quad (5.21)$$

Remark 5.3: If we additionally assume the inverse continuity of $\mathcal{F}$ holds, namely for any $x \in \mathbb{R}^d$ there exists a minimizer x^* of $\mathcal{F}$ (which may depend on x) such that it holds

$$|x - x^*| \le C_0|\mathcal{F}(x) - \underline{\mathcal{F}}|^\ell,$$

where ℓ, C_0 are some positive constants, then one can conclude that $\widetilde{x} \to x^*$ as $\alpha \to \infty$.

Proof: Define

$$T := \inf\left\{t \ge 0 : \mathbb{E}[e^{-\alpha\mathcal{F}(\overline{X}_t)}] < \frac{1}{2}\mathbb{E}[e^{-\alpha\mathcal{F}(\overline{X}_0)}]\right\} \text{ with } \inf\emptyset = \infty. \quad (5.22)$$

Obviously, $T > 0$. Assume that $T < \infty$, then for $t \in [0, T]$, one can deduce that

$$\frac{\lambda\gamma}{2m^2} - \left(\frac{2\lambda^2}{\gamma m} + \frac{\sigma^2}{m^2}\right)\frac{2e^{-\alpha\underline{\mathcal{F}}}}{\mathbb{E}[e^{-\alpha\mathcal{F}(\overline{X}_t)}]}$$

$$\ge \frac{\lambda\gamma}{2m^2} - \left(\frac{2\lambda^2}{\gamma m} + \frac{\sigma^2}{m^2}\right)\frac{4e^{-\alpha\underline{\mathcal{F}}}}{\mathbb{E}[e^{-\alpha\mathcal{F}(\overline{X}_0)}]} = \mu > 0. \quad (5.23)$$

Consequently by (5.10) we have

$$\frac{d}{dt}\mathbb{E}[\mathcal{H}(t)] \le -\frac{\gamma}{m}\mathbb{E}[|\overline{V}_t|^2] - \mu\mathbb{E}[|\delta\overline{X}_t|^2]$$

$$\le -\min\left\{\mu, \frac{\gamma}{m}\right\}(\mathbb{E}[|\delta\overline{X}_t|^2] + \mathbb{E}[|\overline{V}_t|^2])$$

$$\le -\frac{\min\{\mu, \frac{\gamma}{m}\}}{\frac{3}{2}((\frac{\gamma}{2m})^2 + 1)}\mathbb{E}[\mathcal{H}(t)], \quad (5.24)$$

where we have used the estimate (5.1). This implies that

$$\mathbb{E}[\mathcal{H}(t)] \le \mathbb{E}[\mathcal{H}(0)]\exp\left(-\frac{\min\{\mu, \frac{\gamma}{m}\}}{\frac{3}{2}((\frac{\gamma}{2m})^2 + 1)}t\right) = \mathbb{E}[\mathcal{H}(0)]\exp(-\chi t). \quad (5.25)$$

One further notice that

$$\chi \le \frac{\frac{\gamma}{m}}{\frac{3}{2}((\frac{\gamma}{2m})^2 + 1)} < \frac{\gamma}{m}.$$

Set $\mathcal{Y}(t) := (\mathbb{E}[e^{-\alpha\mathcal{F}(\overline{X}_t)}])^2$. Then we have

$$\mathcal{Y}'(0) = -2\alpha\mathbb{E}[e^{-\alpha\mathcal{F}(\overline{X}_0)}]\mathbb{E}[e^{-\alpha\mathcal{F}(\overline{X}_0)}\nabla\mathcal{F}(\overline{X}_0)\cdot\overline{V}_0]. \qquad (5.26)$$

By Gronwall's inequality, it follows from Lemma 5.2 that

$$\frac{d}{dt}\mathcal{Y}(t) \geq \mathcal{Y}'(0)\exp\left(-\frac{\gamma}{m}t\right)$$
$$-4\left(\alpha + \frac{\alpha\lambda}{m}2\left(\frac{2m}{\gamma}\right)^2\right)c_{\mathcal{F}}e^{-2\alpha\mathcal{F}}\exp\left(-\frac{\gamma}{m}t\right)\int_0^t\exp\left(\frac{\gamma}{m}s\right)\mathbb{E}[\mathcal{H}(s)]ds$$
$$\geq \mathcal{Y}'(0)\exp\left(-\frac{\gamma}{m}t\right)$$
$$-4\left(\alpha + \frac{\alpha\lambda}{m}2\left(\frac{2m}{\gamma}\right)^2\right)c_{\mathcal{F}}e^{-2\alpha\mathcal{F}}\mathbb{E}[\mathcal{H}(0)]\exp\left(-\frac{\gamma}{m}t\right)\int_0^t\exp\left(\left(\frac{\gamma}{m}-\chi\right)s\right)ds$$
$$\geq \mathcal{Y}'(0)\exp\left(-\frac{\gamma}{m}t\right)$$
$$-4\left(\alpha + \frac{\alpha\lambda}{m}2\left(\frac{2m}{\gamma}\right)^2\right)c_{\mathcal{F}}e^{-2\alpha\mathcal{F}}\frac{\mathbb{E}[\mathcal{H}(0)]}{\frac{\gamma}{m}-\chi}\exp(-\chi t),$$

which implies that

$$\mathcal{Y}(t) \geq \mathcal{Y}(0) - \frac{m}{\gamma}(-\mathcal{Y}'(0))_+$$
$$-4\left(\alpha + \frac{\alpha\lambda}{m}2\left(\frac{2m}{\gamma}\right)^2\right)c_{\mathcal{F}}e^{-2\alpha\mathcal{F}}\frac{\mathbb{E}[\mathcal{H}(0)]}{\chi(\frac{\gamma}{m}-\chi)}.$$

By assumption (5.20), this means that

$$(\mathbb{E}[e^{-\alpha\mathcal{F}(\overline{X}_t)}])^2$$
$$\geq (\mathbb{E}[e^{-\alpha\mathcal{F}(\overline{X}_0)}])^2 - 2\alpha\frac{\gamma}{m}\mathbb{E}[e^{-\alpha\mathcal{F}(\overline{X}_0)}]\left(\mathbb{E}[e^{-\alpha\mathcal{F}(\overline{X}_0)}\nabla\mathcal{F}(\overline{X}_0)\cdot\overline{V}_0]\right)_+$$
$$-4\left(\alpha + \frac{\alpha\lambda}{m}2\left(\frac{2m}{\gamma}\right)^2\right)c_{\mathcal{F}}e^{-2\alpha\mathcal{F}}\frac{\mathbb{E}[\mathcal{H}(0)]}{\chi(\frac{\gamma}{m}-\chi)}$$
$$\geq \frac{1}{4}(\mathbb{E}[e^{-\alpha\mathcal{F}(\overline{X}_0)}])^2.$$

This means that there exists $\delta > 0$ such that $\mathbb{E}[e^{-\alpha\mathcal{F}(\overline{X}_t)}] \geq \frac{1}{2}\mathbb{E}[e^{-\alpha\mathcal{F}(\overline{X}_0)}]$ in $[T, T+\delta)$ as well. This then contradicts with the definition of T. Hence $T = \infty$. Consequently it holds that

$$\mathbb{E}[\mathcal{H}(t)] \leq \mathbb{E}[\mathcal{H}(0)]\exp(-\chi t) \text{ and } \mathbb{E}[e^{-\alpha\mathcal{F}(\overline{X}_t)}] \geq \frac{1}{2}\mathbb{E}[e^{-\alpha\mathcal{F}(\overline{X}_0)}], \qquad (5.27)$$

for all $t \geq 0$. Recalling the fact (5.9) this infers that

$$\mathbb{E}[|\overline{X}_t - X^\alpha(\rho_t)|^2] \leq 2e^{-\alpha\underline{\mathcal{F}}} \frac{\mathbb{E}[|\delta\overline{X}_t|^2]}{\mathbb{E}[e^{-\alpha\mathcal{F}(\overline{X}_t)}]}$$

$$\leq 4e^{-\alpha\underline{\mathcal{F}}} \left(\frac{2m}{\gamma}\right)^2 \frac{2\mathbb{E}[\mathcal{H}(0)]}{\mathbb{E}[e^{-\alpha\mathcal{F}(\overline{X}_0)}]} \exp(-\chi t). \qquad (5.28)$$

Additionally, one has

$$\mathbb{E}[|\overline{X}_t - \mathbb{E}[\overline{X}_t]|^2] \leq 2\left(\frac{2m}{\gamma}\right)^2 \mathbb{E}[\mathcal{H}(t)]$$

$$\leq C\exp(-\chi t)\mathbb{E}[|\overline{V}_t|^2]$$

$$\leq 2\mathbb{E}[\mathcal{H}(t)] \leq C\exp(-\chi t). \qquad (5.29)$$

Moreover we have

$$\left|\frac{d}{dt}\mathbb{E}[\overline{X}_t]\right| \leq \mathbb{E}[|\overline{V}_t|] \leq C\exp\left(-\frac{1}{2}\chi t\right) \to 0 \text{ as } t \to \infty. \qquad (5.30)$$

This means that $\mathbb{E}[\overline{X}_t] \to \widetilde{x}$ for some $\widetilde{x}$ depending on α, then it follows from (5.29) that $\overline{X}_t \to \widetilde{x}$ in mean square. Thus we have $X^\alpha(\rho_t) \to \widetilde{x}$ according to (5.28). Furthermore, by (5.27) one as $\frac{1}{2}\mathbb{E}[e^{-\alpha\mathcal{F}(\overline{X}_0)}] \leq \mathbb{E}[e^{-\alpha\mathcal{F}(\overline{X}_t)}] \to e^{-\alpha\mathcal{F}(\widetilde{x})}$. Therefore we conclude that

$$\mathcal{F}(\widetilde{x}) \leq \frac{1}{\alpha}\log(2) - \frac{1}{\alpha}\log\left(\mathbb{E}[e^{-\alpha\mathcal{F}(\overline{X}_0)}]\right). \qquad (5.31)$$

By the Laplace principle (3.3), one has

$$0 \leq \mathcal{F}(\widetilde{x}) - \underline{\mathcal{F}} \leq \frac{1}{\alpha}\log(2) - \frac{1}{\alpha}\log(\mathbb{E}[e^{-\alpha\mathcal{F}(\overline{X}_0)}]) - \underline{\mathcal{F}} \to 0 \text{ as } \alpha \to \infty. \qquad (5.32)$$

This completes the proof. □

6. Numerical examples

In this section, we illustrate through various numerical examples the previous theoretical analysis, i.e., the mean-field limit and the small inertial limit, and analyze the performance of SD-PSO-based methods against various prototype global optimization functions. We refer to Refs. [52,57,61,63] for applications of CBO and related methods to high dimensional problems in machine learning.

The SD-PSO algorithm. First we introduce the time discrete versions of the SD-PSO systems [89]. The particle system (2.7) is solved by the *discrete PSO method without local best*

$$
\begin{aligned}
X_i^{n+1} &= X_i^n + \Delta t\, V_i^{n+1}, \\
mV_i^{n+1} &= mV_i^n - \gamma\Delta t\, V_i^{n+1} + \lambda\Delta t\left(\overline{X}_\alpha^n - X_i^n\right) \\
&\quad + \sigma\sqrt{\Delta t}\, D(\overline{X}_\alpha^n - X_i^n)\,\theta_i^n,
\end{aligned}
\tag{6.1}
$$

where $\theta_i \sim \mathcal{N}(0,1)$ and the last equation can be rewritten as

$$
\begin{aligned}
V_i^{n+1} &= \left(\frac{m}{m+\gamma\,\Delta t}\right) V_i^n + \frac{\lambda\,\Delta t}{m+\gamma\,\Delta t}\left(\overline{X}_\alpha^n - X_i^n\right) \\
&\quad + \frac{\sigma\,\sqrt{\Delta t}}{m+\gamma\,\Delta t} D(\overline{X}_\alpha^n - X_i^n)\,\theta_i^n.
\end{aligned}
\tag{6.2}
$$

In the general case, the SD-PSO system (3.33) is solved by the *discrete PSO method with local best*

$$
\begin{aligned}
X_i^{n+1} &= X_i^n + \Delta t\, V_i^{n+1}, \\
V_i^{n+1} &= \left(\frac{m}{m+\gamma\,\Delta t}\right) V_i^n + \frac{\lambda_1\,\Delta t}{m+\gamma\,\Delta t}\,(Y_i^n - X_i^n) \\
&\quad + \frac{\lambda_2\,\Delta t}{m+\gamma\,\Delta t}\left(\overline{Y}_\alpha^n - X_i^n\right) + \frac{\sigma_1\,\sqrt{\Delta t}}{m+\gamma\,\Delta t} D(Y_i^n - X_i^n)\,\theta_{1,i}^n \\
&\quad + \frac{\sigma_2\,\sqrt{\Delta t}}{m+\gamma\,\Delta t} D(\overline{Y}_\alpha^n - X_i^n)\,\theta_{2,i}^n, \\
Y_i^{n+1} &= Y_i^n + \nu\,\Delta t\left(X_i^{n+1} - Y_i^n\right) S^\beta(X_i^{n+1}, Y_i^n),
\end{aligned}
\tag{6.3}
$$

where $\theta_{1,i}, \theta_{2,i} \sim \mathcal{N}(0,1)$.

Remark 6.1: Note that, the numerical scheme (6.3) using uniform noise becomes equivalent to the PSO algorithm (2.12) under assumptions (2.8) for $\Delta t = 1$, $\nu = 0.5$, and taking the limit $\alpha, \beta \to \infty$ so that $Y_i^n, \overline{Y}_\alpha^n$ match the local and global best definitions in (2.3). In addition, in the limit $m \to 0^+$ scheme (6.3) is consistent with the zero-inertia limit (4.35) and reduces to the *discrete CBO method with local best*

$$
X_i^{n+1} = X_i^n + \lambda_1 \, \Delta t \, (Y_i^n - X_i^n) + \lambda_2 \, \Delta t \left(\overline{Y}_\alpha^n - X_i^n \right)
$$
$$
+ \sigma_1 \, \sqrt{\Delta t} D(Y_i^n - X_i^n) \, \theta_{1,i}^n + \sigma_2 \, \sqrt{\Delta t} D(\overline{Y}_\alpha^n - X_i^n) \, \theta_{2,i}^n, \qquad (6.4)
$$
$$
Y_i^{n+1} = Y_i^n + \nu \, \Delta t \left(X_i^{n+1} - Y_i^n \right) S^\beta(X_i^{n+1}, Y_i^n).
$$

6.1. *Validation of the mean-field limit*

In the following we validate numerically the mean-field limit by considering as prototype functions for global optimization the Ackley function and the Rastrigin function in one dimension. The functions have multiple local minima that can easily trap the particle dynamics (see Fig. 3). We refer to Ref. [68] for additional examples.

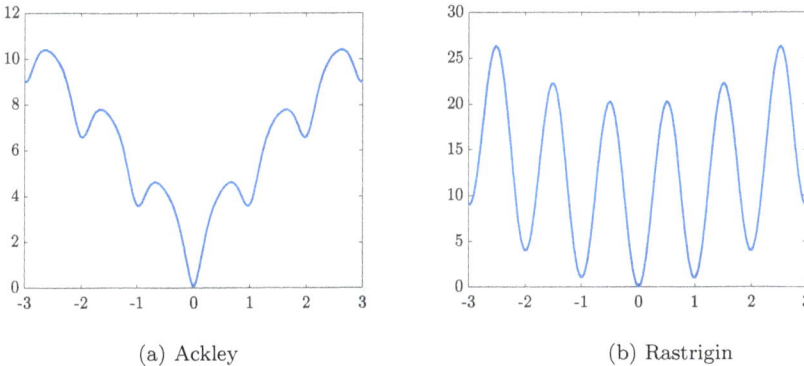

(a) Ackley   (b) Rastrigin

Fig. 3. One-dimensional Ackley and Rastrigin functions in the interval $[-3, 3]$ with global minimum in the origin.

The MF-PSO solver. The corresponding MF-PSO equation without local best (3.4) has been discretized using a dimensional splitting where the transport part is solved through a backward semi-Lagrangian method and the remaining Fokker-Planck term is discretized using an implicit central scheme. The MF-PSO equation with memory (3.36) is solved by a further dimensional splitting where the additional memory term is discretized using a Lax-Wendroff method. Zero boundary conditions have been implemented outside the computational domain. We refer Refs. [68,90] for further details and additional discretizations of Vlasov-Fokker-Planck systems.

In the sequel we used $N = 5 \times 10^5$ particles, a mesh size for the mean-field solver of 90×120 points for $(x, v) \in [-3, 3] \times [-4, 4]$, and whenever present, the mesh and domain size in y have been taken identical to those in x. To represent the particle solution, we used the probability density estimate based on a normal kernel reconstruction evaluated at equally-spaced points. In all simulations, the initial distribution is assumed to be uniform and the minimum is assumed in $x = 0$.

6.1.1. *Absence of memory effects*

We consider the optimization process of the Ackley function. Here we report the results obtained with

$$\gamma = 0.5, \quad \lambda = 1, \quad \sigma = 1/\sqrt{3}, \quad \alpha = 30. \tag{6.5}$$

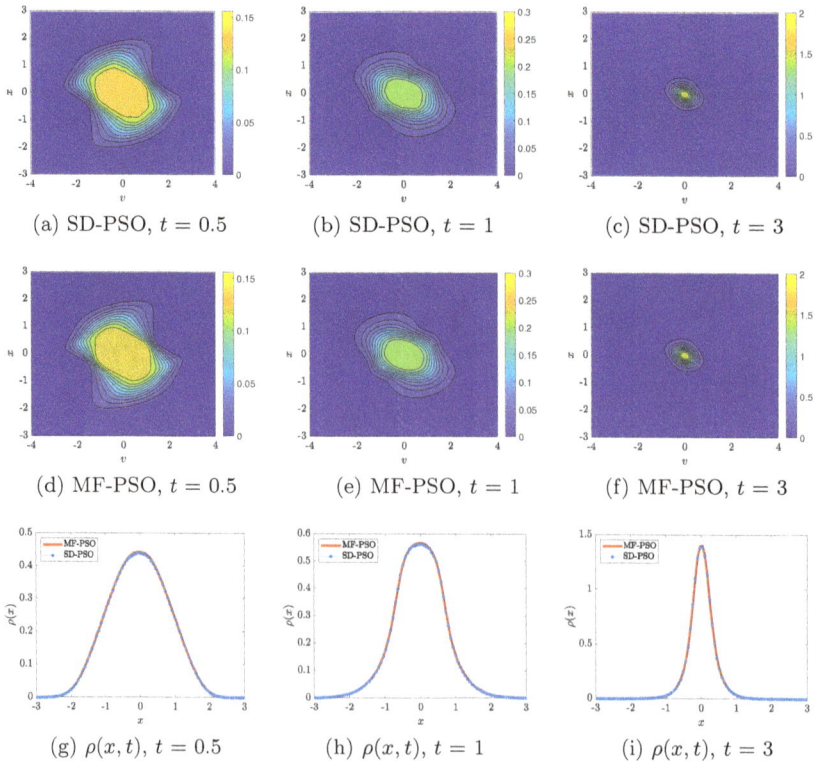

(a) SD-PSO, $t = 0.5$ (b) SD-PSO, $t = 1$ (c) SD-PSO, $t = 3$

(d) MF-PSO, $t = 0.5$ (e) MF-PSO, $t = 1$ (f) MF-PSO, $t = 3$

(g) $\rho(x, t)$, $t = 0.5$ (h) $\rho(x, t)$, $t = 1$ (i) $\rho(x, t)$, $t = 3$

Fig. 4. Mean-field validation (no memory). Optimization of the Ackley function. First row: solution of the SD-PSO system (3.1) using $N = 5 \times 10^5$ particles. Second row: solution of the MF-PSO limit (3.4). Third row: marginal densities.

The values of λ and σ correspond to the standard PSO choice $c_k = 2$ in (2.8). In Fig. 4 we report the contour plots of the evolution, at times $t = 0.5$, $t = 1$ and $t = 3$, of the particle distribution computed through (6.1) and by the direct discretization of the mean-field Eq. (3.4) together with the evolution in time of the marginal density $\rho(x,t) = \int_{\mathbb{R}^d} f(x,v,t)\,dv$.

6.1.2. *Only local best dynamics*

In the second test case we introduce the dependence from the memory variable and compare the solutions of the discretized stochastic particle model (6.3) with the solver of the mean-field limit (3.36) in the case of the Rastrigin function. We assume $\lambda_2 = 0$ and $\sigma_2 = 0$, i.e. only the local best is present. The same parameters (6.5) have been used together with $\beta = 30$ and $\nu = 0.5$ for the local best. In Fig. 5 we report the contour plot of

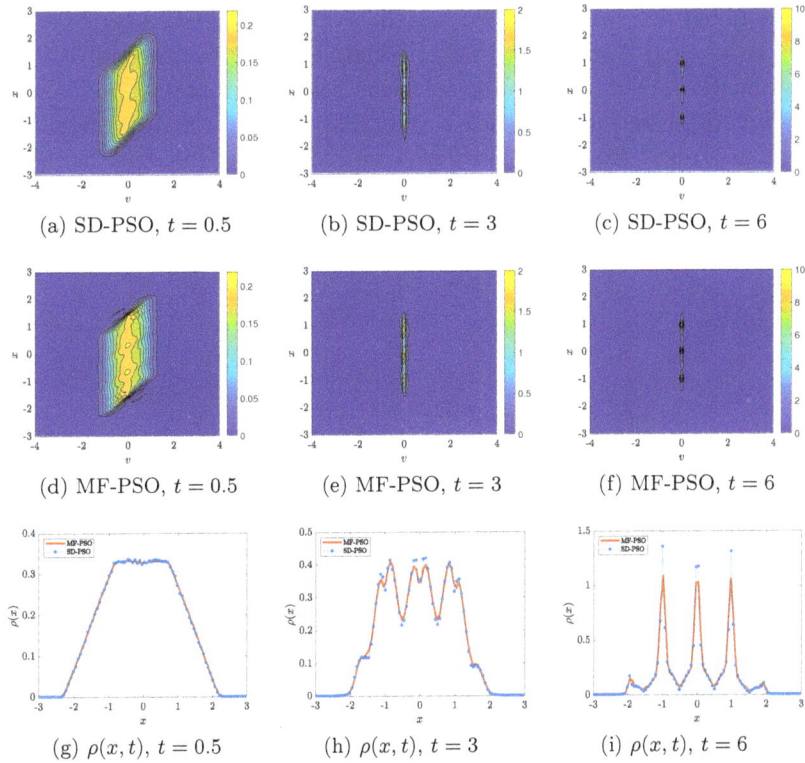

(a) SD-PSO, $t = 0.5$ (b) SD-PSO, $t = 3$ (c) SD-PSO, $t = 6$

(d) MF-PSO, $t = 0.5$ (e) MF-PSO, $t = 3$ (f) MF-PSO, $t = 6$

(g) $\rho(x,t)$, $t = 0.5$ (h) $\rho(x,t)$, $t = 3$ (i) $\rho(x,t)$, $t = 6$

Fig. 5.   Mean-field validation (local best only). Optimization of the Rastrigin function with minimum in $x = 0$. First row: solution of the SD-PSO system (3.33). Second row: solution of the MF-PSO limit (3.36). Third row: marginal densities.

the particle and mean-field solutions for the Rastrigin function, where now
the final simulation time is $t = 6$. The corresponding marginal densities
are also reported. Also in this second case, one can appreciate the good
agreement between the particle and mean-field solutions. We can note that
in the presence of local best only, the particles tend to return to their local
best position creating a "memory effect" that leads them to concentrate
not only in the global minimum but also in the local minima. For large
times we obtain a sequence of particle peaks with zero speed exactly in the
positions of the local minima. Thus the dynamic allows us to identify each
type of minimum present in the functions.

6.1.3. The general case

In the final test case, we keep the previous scenario, adding the contribution
of the global best with the same weight as the local best. Therefore, we
take $\lambda_1 = \lambda_2 = 1$, $\sigma_1 = \sigma_2 = 1/\sqrt{3}$ and the same parameters (6.5) in
our numerical experiments. The solutions have been obtained by solving
the discretized stochastic particle system (6.3) and the deterministic solver
of the mean-field Eq. (3.36). In Fig. 6 we report the associated marginal
density plots. One can observe that the local minima effect disappears and

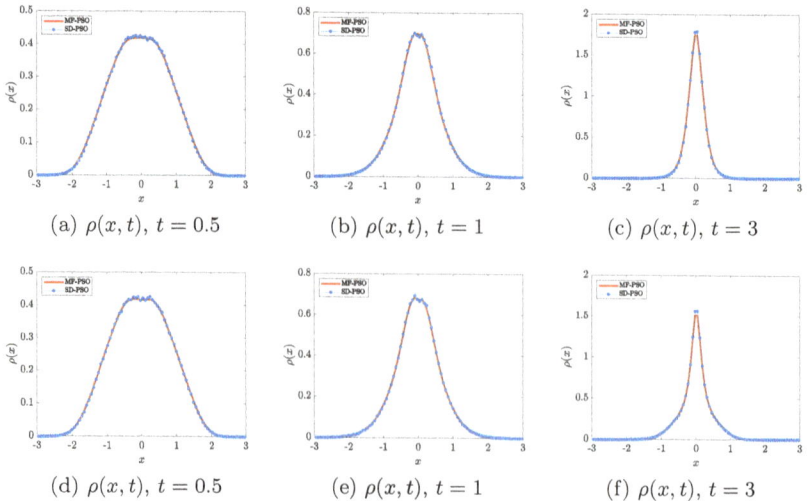

(a) $\rho(x,t)$, $t = 0.5$ \qquad (b) $\rho(x,t)$, $t = 1$ \qquad (c) $\rho(x,t)$, $t = 3$

(d) $\rho(x,t)$, $t = 0.5$ \qquad (e) $\rho(x,t)$, $t = 1$ \qquad (f) $\rho(x,t)$, $t = 3$

Fig. 6.   Mean-field validation (general case). Evolution of the density $\rho(x,t)$ of the SD-
PSO system (3.33) and the MF-PSO limit (3.36) for two different one-dimensional function with minimum in $x = 0$. First row: optimization on the Ackley function. Second
row: optimization oh the Rastrigin function.

the systems converge consistently towards the global minimum. Note that, by comparing the results for the Ackley function in Fig. 6 and those in the last row of Fig. 4 obtained by solving the same problem in absence of memory terms, at the same time instants, a faster convergence towards the global minimum is observed.

6.2. *Numerical small inertia limit*

From the analysis in Sec. 4, the classical CBO model (4.7) is produced as a hydrodynamic approximation of the mean-field PSO system (3.4) in the limit of small inertia. Therefore, we compare the particle solution to a discretization of the mean-field limit CBO system (4.7), starting from the discretization of the stochastic particle model without memory effect (6.1) and decreasing the inertial weight $m \to 0$ ($\gamma \to 1$).

In Fig. 7 we report the plots of the density that describes the solution of the mean-field CBO model and the stochastic PSO model for different inertial weights ($m = 0.5$, $m = 0.1$ and $m = 0.01$). We considered the minimization problem for the Ackley function with minimum in $x = 0$ and in $x = 1$ with $N = 5 \times 10^5$ particles for the SD-PSO discretization and a grid of 120 points in space for the mean-field CBO solver. It is clear that in the case of $m = 0.5$ the two densities at the final time $t = 2$ are considerably

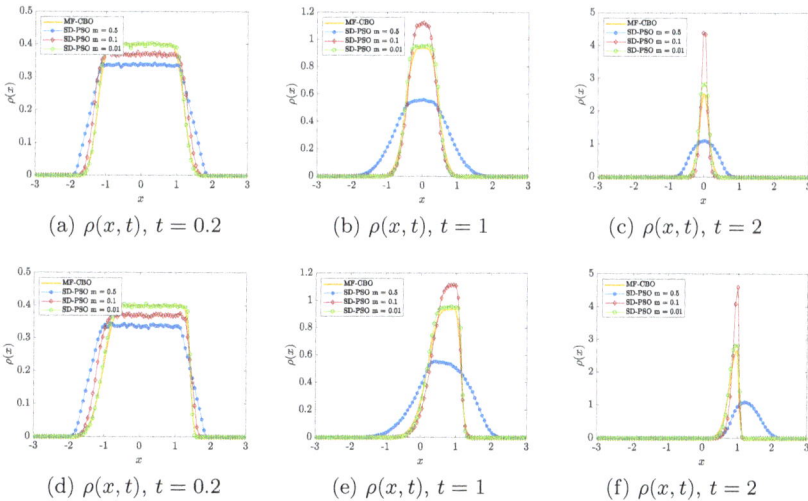

(a) $\rho(x,t)$, $t = 0.2$ (b) $\rho(x,t)$, $t = 1$ (c) $\rho(x,t)$, $t = 2$

(d) $\rho(x,t)$, $t = 0.2$ (e) $\rho(x,t)$, $t = 1$ (f) $\rho(x,t)$, $t = 2$

Fig. 7. Low inertia limit. Evolution of the density $\rho(x,t)$ of the SD-PSO discretization (6.1), for decreasing inertial weight $m = 0.5, 0.1, 0.01$, and the mean-field CBO model (4.7) for the Ackley function with a uniform initial data. First row: minimum in $x = 0$. Second row: minimum in $x = 1$.

different and a slower convergence is observed in the SD-PSO system, for $m = 0.1$ the agreement is higher and the particle solution seems to converge faster to the minimum, finally in the case $m = 0.01$ both densities simultaneously grow towards a Dirac delta centered in the minimum. For smaller values of m the two solutions becomes indistinguishable and we omitted the results.

6.3. Performance on high-dimensional test cases

In this section we report the results of several experiments concerning the behavior of the stochastic PSO models, discretized using (6.1) in absence of memory or (6.3) in the general case, in high dimension ($d = 20$) for various prototype test functions (see Table 5). Defining the success rate is critical as it completely alters the performance of the algorithm. In particular, depending on the shape of the objective function, the distance between the estimated minimum and the real minimum can be used as an indicator as in Refs. [50–52, 60, 61]. For some functions, however, this choice may be a poor indicator of the algorithm's performance, since the corresponding value function may be far from its minimum optimal value. In the first round of test cases, since we are limited to the Ackley and Rastrigin functions, for comparison purposes we rely on the choice reported below. Later, when we test the performance of the algorithm for a broader spectrum of test functions, we will generalize the definition of success rate by including the value of the function [53].

Thus we define:

- the *success rate*, computed averaging over n_r runs and using as convergence criterion

$$\|\overline{X}_\alpha^{n_*} - x^*\|_\infty < \delta_{err}, \qquad \text{or} \qquad \|\overline{Y}_\alpha^{n_*} - x^*\|_\infty < \delta_{err}$$

where x^* is the minimum and n_* the final time.
- the *error*, evaluated as expected value in the L_2 norm over the successful runs

$$\mathbb{E}(\|\overline{X}_\alpha^{n_*} - x^*\|_2), \qquad \text{or} \qquad \mathbb{E}(\|\overline{Y}_\alpha^{n_*} - x^*\|_2);$$

- the *number of iterations*, where we stop the iteration if

$$\|\overline{X}_\alpha^n - \overline{X}_\alpha^{n-1}\| < \delta_{stall}, \qquad \text{or} \qquad \|\overline{Y}_\alpha^n - \overline{X}_\alpha^{n-1}\| < \delta_{stall}$$

for n_{stall} consecutive iterations or a maximum n_{max} iterations has been reached.

In the sequel, we consider $n_r = 500$, $\delta_{err} = 0.25$, $\delta_{stall} = 10^{-4}$ and $n_{max} = 10^4$. We remark that, increasing the problem dimension, a larger value of $\alpha \gg 1$ provides better performance [50, 61]. On the other hand, a large value of α may generate numerical instabilities given by the definition of the regularized global best. To avoid this, we used the algorithm presented in Ref. [61] which allows the use of arbitrary large values of α.

In the following test cases, we address the role of the various parameters, of the presence of memory and of the local best when solving high dimensional global optimization problems. We refer also to Ref. [54] for additional comparisons. In our experiments, the PSO constraints (2.8) have shown strong limitations in terms of success rates and have not been considered. We refer to Ref. [68] for further details and comparisons.

6.3.1. *Effect of the inertial parameter m*

First we test the algorithm performance for the Ackley and the Rastrigin functions in $[-3, 3]^d$, $d = 20$. In the left column of Table 1 and Table 2 we report the results obtained without memory effects (6.1) and in the right column the results with memory effects (6.3). Since, typically, optimizing the Rastrigin function is far more difficult than the Ackley function, we explore the space of parameters searching for optimal values of σ and Δt for the Rastrigin function, then we used the same values for the Ackley function. This optimization was done empirically through several simulations with simple variations of a given step size for the parameters.

Table 1. SD-PSO with and without memory for $\lambda_1 = \sigma_1 = 0$, $\lambda_2 = 1$, $\Delta t = 0.01$, $\nu = 50$, $\beta = 3 \times 10^3$ and $\alpha = 5 \times 10^4$.

Rastrigin			Case without memory				Case with memory		
m		σ	$N = 50$	$N = 100$	$N = 200$	σ_2	$N = 50$	$N = 100$	$N = 200$
0.00	Rate	9.0	100.0%	100.0%	100.0%	11.0	100.0%	100.0%	100.0%
	Error		1.19e-04	1.11e-04	9.68e-05		6.83e-04	4.70e-04	4.69e-04
	n_{iter}		10000.0	10000.0	9912.4		10000.0	9878.2	3290.2
0.01	Rate	7.0	100.0%	100.0%	100.0%	9.0	100.0%	100.0%	100.0%
	Error		9.74e-05	2.01e-05	1.62e-05		8.60e-04	8.56e-04	8.81e-04
	n_{iter}		10000.0	6899.2	2060.1		9939.5	7012.2	5422.1
0.05	Rate	3.5	37.0%	74.0%	94.0%	4.5	100.0%	100.0%	100.0%
	Error		4.27e-04	1.26e-04	1.14e-04		1.15e-03	6.67e-04	6.54e-04
	n_{iter}		8233.2	7814.0	7326.6		9978.0	7657.6	5639.7
0.10	Rate	2.0	1.0%	5.5%	29.5%	3.0	80.8%	96.8%	100.0%
	Error		2.00e-04	1.28e-04	1.11e-04		2.94e-03	8.96e-04	8.24e-04
	n_{iter}		6155.4	6221.9	6214.3		9661.5	8676.5	7331.8

Table 2. SD-PSO with and without memory for $\lambda_1 = \sigma_1 = 0$, $\lambda = \lambda_2 = 1$, $\Delta t = 0.01$, $\nu = 50$, $\beta = 3 \times 10^3$ and $\alpha = 5 \times 10^4$.

Ackley			Case without memory				Case with memory		
m		σ	$N = 50$	$N = 100$	$N = 200$	σ_2	$N = 50$	$N = 100$	$N = 200$
0.00	Rate	9.0	100.0%	100.0%	100.0%	11.0	100.0%	100.0%	100.0%
	Error		8.46e-05	4.20e-05	1.27e-05		1.02e-04	7.66e-05	5.44e-05
	n_{iter}		1364.9	1032.4	869.2		2457.0	1778.0	1513.1
0.01	Rate	7.0	100.0%	100.0%	100.0%	9.0	100.0%	100.0%	100.0%
	Error		9.49e-05	5.89e-05	2.81e-05		2.34e-03	1.91e-04	1.61e-04
	n_{iter}		2192.9	1886.7	1723.6		6430.4	5447.8	4598.3
0.05	Rate	3.5	100.0%	100.0%	100.0%	4.5	100.0%	100.0%	100.0%
	Error		2.27e-04	1.48e-04	1.03e-04		2.41e-04	1.84e-04	1.48e-04
	n_{iter}		5367.3	4459.4	3928.4		7186.1	5996.0	5074.6
0.10	Rate	2.0	99.5%	100.0%	100.0%	3.0	100.0%	100.0%	100.0%
	Error		8.31e-04	2.76e-04	1.91e-04		3.90e-03	2.64e-03	2.06e-03
	n_{iter}		5480.8	4514.1	3909.4		8590.6	7326.4	6350.2

The results are given for different numbers of particles N. We consider $\alpha = 5 \times 10^4$, whereas the memory parameters β and ν were chosen respectively $\beta = 3 \times 10^3$ and $\nu = 50$. Note that, even if we rely only on the global best since we fix $\lambda_1 = \sigma_1 = 0$, due to the regularization of the memory process the two approaches, with and without memory, differs and a higher noise in required in presence of memory. Low inertia values yields better performances overall, however, it should be noticed that for the Rastrigin function the best results in term of convergence speed are obtained with a small but non zero inertia value of $m = 0.01$.

6.3.2. Effect of the local best dynamics

Subsequently, we have introduced the local best dynamics in the same optimization process. To reduce the number of free parameters we assume $\lambda_1 = \xi \cdot \lambda_2$, $\sigma = \xi \cdot \sigma_2$ with $\xi \in [0,1]$ so that the local best is always weighted less than the global best. In this test we keep the inertial value $m = 0$ and $\lambda_1 = 1$, so that we are solving the generalized stochastic differential CBO model with memory using algorithm (6.4). For each value of ξ reported, we have computed an optimal σ_2 achieving the maximum rate of success. We chose $\beta = 3 \times 10^3$, $\Delta t = 0.01$, $\nu = 50$ and $\alpha = 5 \times 10^4$ as in the previous case.

In Tables 3 and 4 we report the behavior of the particle optimizer on the Ackley and Rastrigin functions for different positions of the minimum $x^* = 0$, $x^* = 1$ and $x^* = 2$. Since for large values of ξ we must decrease σ_2 to achieve maximum convergence rate we observe that the total number of

Table 3.   SD-PSO with memory $(m = 0)$ for $\lambda_1 = \xi \cdot \lambda_2$, $\sigma = \xi \cdot \sigma_2$, λ_1, $\lambda_2 = 1$, $\Delta t = 0.01$, $\nu = 50$, $\beta = 3 \times 10^3$, $\alpha = 5 \times 10^4$.

Rastrigin		Case $\xi = 0$, $\sigma_2 = 11.0$			Case $\xi = 0.25$, $\sigma_2 = 8.5$		
		$N = 50$	$N = 100$	$N = 200$	$N = 50$	$N = 100$	$N = 200$
$x^* = 0$	Rate	100.0%	100.0%	100.0%	100.0%	100.0%	100.0%
	Error	7.04e-04	4.58e-04	3.29e-04	9.28e-04	6.11e-04	4.31e-04
	n_{iter}	10000.0	9963.9	4635.1	9978.0	8311.5	5754.1
$x^* = 1$	Rate	98.8%	100.0%	100.0%	99.2%	100.0%	100.0%
	Error	7.08e-04	4.60e-04	3.27e-04	9.31e-04	6.74e-04	4.59e-04
	n_{iter}	10000.0	10000.0	4670.0	9987.0	9746.7	7460.1
$x^* = 2$	Rate	96.0%	99.1%	100.0%	93.5%	100.0%	100.0%
	Error	6.91e-04	4.52e-04	3.28e-04	8.78e-04	6.74e-04	5.66e-04
	n_{iter}	10000.0	10000.0	5035.5	9980.3	9854.1	8971.9

Table 4.   SD-PSO with memory $(m = 0)$ for $\lambda_1 = \xi \cdot \lambda_2$, $\sigma = \xi \cdot \sigma_2$, $\lambda_2 = 1$, $\Delta t = 0.01$, $\nu = 50$, $\beta = 3 \times 10^3$, $\alpha = 5 \times 10^4$.

Ackley		Case $\xi = 0$, $\sigma_2 = 11.0$			Case $\xi = 0.25$, $\sigma_2 = 8.5$		
		$N = 50$	$N = 100$	$N = 200$	$N = 50$	$N = 100$	$N = 200$
$x^* = 0$	Rate	100.0%	100.0%	100.0%	100.0%	100.0%	100.0%
	Error	7.36e-05	5.13e-05	3.26e-05	2.54e-05	1.13e-05	1.07e-05
	n_{iter}	2778.6	2030.0	1623.0	1942.9	1663.8	1442.5
$x^* = 1$	Rate	100.0%	100.0%	100.0%	100.0%	100.0%	100.0%
	Error	7.31e-05	5.14e-05	3.26e-05	2.58e-05	1.12e-05	1.02e-05
	n_{iter}	5298.5	3640.6	2575.9	2465.3	1948.5	1632.5
$x^* = 2$	Rate	100.0%	100.0%	100.0%	100.0%	100.0%	100.0%
	Error	7.30e-05	5.07e-05	3.22e-05	2.64e-05	1.09e-05	1.01e-05
	n_{iter}	7819.8	5771.3	4235.9	3126.8	2286.0	1803.8

iterations may decrease and that a speed-up is obtained thanks to the local best.

6.3.3. *Comparison on prototype functions*

In the last test case we analyze the performance of the methods by solving simultaneously a set of different optimization functions considered in their standard search domains [91] (see Table 5). Here, instead of trying to find an optimal set of parameters for each function we use the same parameters for all functions. Furthermore, in order to identify a comparable set of optimization parameters for the different functions, we found it particularly effective to rescale all functions from their classical domain to the same reference domain. In our experiment we generalized the notion of success

Table 5.   Prototype test functions for global optimization.

Name	Function $\mathcal{F}(x)$	Range	x^*	$\mathcal{F}(x^*)$	Sketch in 2D
Ackley	$-20\exp\left(-0.2\sqrt{\frac{1}{d}\sum_{i=1}^{d}(x_i)^2}\right) - \exp\left(\frac{1}{d}\sum_{i=1}^{d}\cos\left(2\pi(x_i)\right)\right) + 20 + e$	$[-32,32]^d$	$(0,\dots,0)$	0	
Griewank	$1 + \sum_{i=1}^{d}\frac{(x_i)^2}{4000} - \prod_{i=1}^{d}\cos\left(\frac{x_i}{i}\right)$	$[-600,600]^d$	$(0,\dots,0)$	0	
Rastrigin	$10d + \sum_{i=1}^{d}\left[(x_i)^2 - 10\cos\left(2\pi(x_i)\right)\right]$	$[-5.12,5.12]^d$	$(0,\dots,0)$	0	
Rosenbrock	$1 - \cos\left(2\pi\sqrt{\sum_{i=1}^{d}(x_i)^2}\right) + 0.1\sqrt{\sum_{i=1}^{d}(x_i)^2}$	$[-5,10]^d$	$(1,\dots,1)$	0	
Salomon	$1 - \cos\left(2\pi\sqrt{\sum_{i=1}^{d}(x_i)^2}\right) + 0.1\sqrt{\sum_{i=1}^{d}(x_i)^2}$	$[-100,100]^d$	$(0,\dots,0)$	0	
Schwefel 2.20	$\sum_{i=1}^{d}|x_i|$	$[-100,100]^d$	$(0,\dots,0)$	0	
XSY random	$\sum_{i=1}^{d}\eta_i|x_i|^i, \quad \eta_i \sim \mathcal{U}(0,1)$	$[-5,5]^d$	$(0,\dots,0)$	0	
XSY 4	$\left(\sum_{i=1}^{d}\sin^2(x_i) - e^{-\sum_{i=1}^{d}(x_i)^2}\right)e^{-\sum_{i=1}^{d}\sin^2\sqrt{|x_i|}}$	$[-10,10]^d$	$(0,\dots,0)$	-1	

criteria by introducing the following definitions

- the *success rate*, computed averaging over n_r runs and using as convergence criterion

$$\|\overline{Y}_\alpha^{n_*} - x^*\|_\infty < \delta_{err} \quad \text{or} \quad |\mathcal{F}(\overline{Y}_\alpha^{n_*}) - \mathcal{F}(x^*)| < \delta_{fun}$$

  where x^* is the minimum and n_* the final time.
- The *average function value* $\mathcal{F}_{avg}$, computed averaging the function value $\mathcal{F}(\overline{Y}_\alpha^{n_*})$ over n_r runs.

In our simulations, we set $[-1, 1]^d$ as the reference domain and translate the functions so that all have a minimum value of $\mathcal{F}(x^*) = 0$. We selected $\delta_{err} = 0.1$, $\delta_{fun} = 0.01$, $n_r = 500$ and $n_{max} = 10^4$. We let most parameters fixed as in previous test case, namely $\alpha = 5 \times 10^4$, $\beta = 3 \times 10^3$, $\nu = 50$. Additionally we keep $m = 0$, $\Delta t = 0.01$, and for a given value of $\xi = 0$ (absence of local best) and $\xi = 0.25$ (local best weighted $1/4$ of global best) estimate the value for σ_2 in order to maximize the average convergence rate among all functions. This has been done with simple variations of step 0.5 for σ_2 in the simulations, according to results in Fig. 8 where we considered the behavior of the average error and fitness value for different values of σ_2 calculated over n_r runs.

The results in Table 6 confirm the potential of the method in identifying correctly the global minima for different heterogeneous test functions. Overall, with the exception of the Rastrigin function for which the local best produces a reduction in the convergence rate using this set of parameters, the importance of the local best is evident. In particular, the presence of the local best yields a reduction in the number of iterations for the Griewank, the Rosenbrock and the Salomon functions and an increase in the convergence rate for the XSY random and XSY4 functions. Except for the Griewank and Solomon functions, the computed value of the objective function is consistently close to zero and improves by increasing the number of particles. Finally, we emphasize that it was beyond the scope of this chapter to discuss additional practical improvements to the algorithms that can be adopted to improve the success rate and the efficiency, like the use of random batch methods [52, 79, 80], particle reduction techniques [61, 62] and parameters adaptivity [4, 7]. We refer to Ref. [67] for further details on these implementation aspects.

(a) $\|\overline{Y}_\alpha^{n*} - x^*\|_\infty$ with $\xi = 0$

(b) $\|\overline{Y}_\alpha^{n*} - x^*\|_\infty$ with $\xi = 0.25$

(c) $\mathcal{F}_{avg}$ with $\xi = 0$

(d) $\mathcal{F}_{avg}$ with $\xi = 0.25$

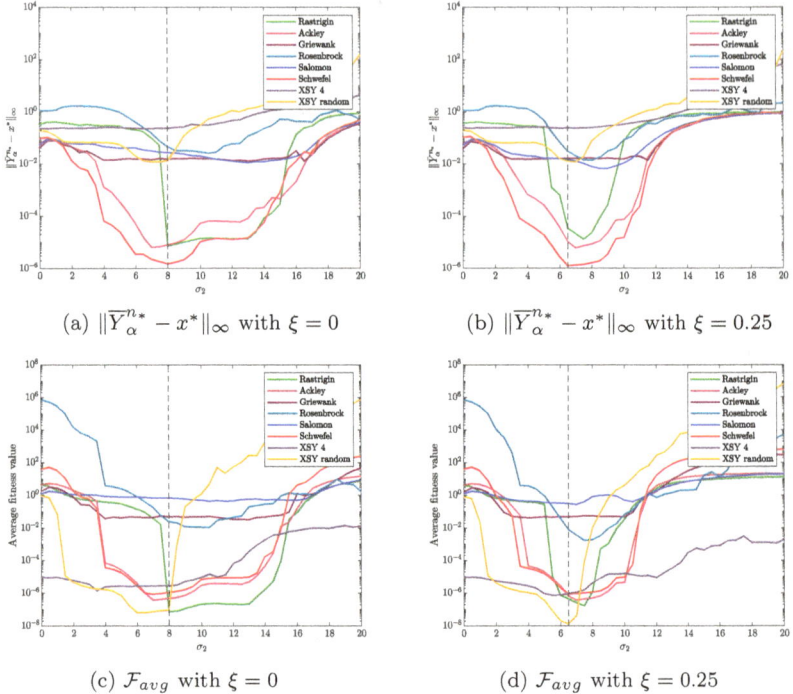

Fig. 8. SD-PSO with memory ($m = 0$). Behavior of the average error (top) and fitness value (bottom) for different values of σ_2. Here $\sigma_1 = \xi \cdot \sigma_2$, $\lambda_1 = \xi \cdot \lambda_2$, $\lambda_2 = 1$, $\Delta t = 0.01$, $\nu = 50$, $\beta = 3 \times 10^3$ and $\alpha = 5 \times 10^4$. The dashed vertical lines are the estimated optimal values.

7. Concluding remarks and research directions

PSO methods represent a particularly prominent category within global optimization methods that do not make use of the gradient of the objective function. The popularity of these methods is related to the versatility and robustness of the algorithms, the good scalability that allows dealing with high-dimensional problems, and the ability to identify the global minimum effectively even in the case of non-convex and possibly non-smooth functions. Despite this, a complete mathematical theory related to the derivation of such methods and their global convergence properties is still lacking.

In this chapter, relying on some recent results [67, 68, 70–72], we have made an important step towards the construction of a general mathematical theory for the rigorous analysis and the understanding of PSO methods. The starting point of our analysis is a generalization of PSO methods in the context of second-order stochastic differential equations. In addition

Table 6. SD-PSO with memory ($m = 0$) for $\lambda_1 = \xi \cdot \lambda_2$, $\sigma_1 = \xi \cdot \sigma_2$, $\lambda_2 = 1$, $\Delta t = 0.01$, $\nu = 50$, $\beta = 3 \times 10^3$, $\alpha = 5 \times 10^4$.

		Case $\xi = 0$, $\sigma_2 = 8.0$			Case $\xi = 0.25$, $\sigma_2 = 6.5$		
		$N = 50$	$N = 100$	$N = 200$	$N = 50$	$N = 100$	$N = 200$
Ackley	Rate	100.0%	100.0%	100.0%	100.0%	100.0%	100.0%
	Error	9.44e-05	3.57e-05	1.48e-05	9.25e-06	4.40e-06	2.02e-06
	$\mathcal{F}_{avg}$	2.61e-05	1.04e-05	8.49e-06	2.65e-05	1.26e-05	5.78e-06
	n_{iter}	1012.5	847.9	736.2	1033.4	874.3	764.0
Griewank	Rate	100.0%	100.0%	100.0%	100.0%	100.0%	100.0%
	Error	2.28e-02	2.24e-02	2.19e-02	2.27e-02	2.16e-02	2.24e-02
	$\mathcal{F}_{avg}$	5.57e-02	5.21e-02	4.26e-02	5.25e-02	4.93e-02	2.28e-02
	n_{iter}	1010.8	861.6	761.7	1006.3	734.7	626.6
Rastrigin	Rate	34.0%	70.7%	95.0%	9.0%	26.4%	42.0%
	Error	1.78e-05	1.89e-05	2.05e-05	3.01e-05	3.12e-05	3.03e-05
	$\mathcal{F}_{avg}$	9.32e-08	9.68e-08	9.95e-08	2.41e-07	2.58e-07	2.44e-07
	n_{iter}	1308.5	1122.9	970.5	1631.0	1483.0	1334.8
Rosenbrock	Rate	49.3%	84.7%	100.0%	87.3%	100.0%	100.0%
	Error	2.60e-02	3.44e-02	1.08e-02	4.87e-02	3.32e-02	6.92e-03
	$\mathcal{F}_{avg}$	8.58e-02	1.25e-02	9.30e-03	2.12e-02	8.01e-03	3.23e-04
	n_{iter}	8009.3	8392.8	7358.0	9669.8	9553.8	7925.7
Schwefel 2.20	Rate	100.0%	100.0%	100.0%	100.0%	100.0%	100.0%
	Error	2.11e-05	1.73e-06	7.32e-07	3.65e-06	1.63e-06	1.09e-06
	$\mathcal{F}_{avg}$	2.93e-03	4.99e-04	2.18e-04	5.14e-05	2.46e-05	8.01e-06
	n_{iter}	865.9	749.8	668.3	863.2	747.0	665.8
Salomon	Rate	84.7%	98.7%	100.0%	100.0%	100.0%	100.0%
	Error	8.94e-02	6.45e-02	4.99e-02	3.72e-02	3.21e-02	2.75e-02
	$\mathcal{F}_{avg}$	8.96e-01	6.66e-01	5.24e-01	3.83e-01	3.21e-01	2.75e-01
	n_{iter}	1749.3	1657.9	1631.9	2193.7	1749.7	1138.2
XSY random	Rate	90.0%	99.3%	100.0%	100.0%	100.0%	100.0%
	Error	4.11e-02	2.26e-02	1.14e-02	2.45e-02	1.67e-02	1.66e-02
	$\mathcal{F}_{avg}$	5.64e-07	9.60e-08	6.06e-08	9.75e-09	7.26e-09	4.56e-09
	n_{iter}	10000.0	10000.0	10000.0	10000.0	10000.0	10000.0
XSY 4	Rate	100.0%	100.0%	100.0%	100.0%	100.0%	100.0%
	Error	1.09e+00	9.85e-01	9.70e-01	8.56e-01	8.19e-01	7.97e-01
	$\mathcal{F}_{avg}$	2.88e-05	2.57e-05	7.44e-05	1.69e-07	1.42e-07	1.41e-07
	n_{iter}	9682.5	9018.1	8861.6	10000.0	10000.0	10000.0

to the continuous formulation of PSO algorithms this novel class of methods generalizes the particle optimization process by making the alignment and exploration coefficients, based on the corresponding drift and diffusion dynamics, independent.

In the mean-field limit, using a regularized version of these SD-PSO systems, we obtained a Vlasov-Fokker-Planck type equation describing the MF-PSO dynamics. In addition, we rigorously studied the behavior of the system for small values of the inertia parameter showing how in such a limit the MF-PSO dynamics converges to a generalization of CBO models containing the local best. The latter result allowed us to clarify the relationships between these two classes of meta-heuristic optimization methods. A convergence result to the global minimum for a wide class of objective function is then proved in the case where the dynamic does not take into account memory effects. A complete gallery of numerical examples illustrate on the one hand the theoretical results obtained and on the other hand how the new class of SD-PSO methods potentially presents several advantages over traditional PSO in terms of convergence speed and solution stability.

These results open important perspectives in the area of mathematical understanding of particle swarming optimization methods and in the construction of new algorithms. Among the many research directions some, not exhaustive, are summarized below.

- The majority of PSO applications are limited to single objective and unconstrained optimization problems. Therefore, the development of methods capable to deal with multi-objective and constrained optimization problems is a challenging and interesting area of research.

- Most of the convergence results for mean-field PSO and CBO models refer to the global best only. Generalization of these results to include the effect of the local best and its role should be studied. Convergence rate estimates of practical interest are still limited and further analysis is necessary.

- Similarly to classical PSO algorithms, the computational parameters are usually determined according to specific problems and require considerable application experience and numerous experimental tests. The identification of optimal parameters and the implementation of adaptive techniques for their determination is thus fundamental for many applications.

Acknowledgments

This chapter has been written within the activities of GNCS group of INdAM (National Institute of High Mathematics). The support of MIUR-PRIN Project 2017, No. 2017KKJP4X "Innovative numerical methods for evolutionary partial differential equations and applications" and of the ESF PhD grant "Mathematical and statistical methods for machine learning in biomedical and socio-sanitary applications" is acknowledged. H. H. is partially supported by the Pacific Institute for the Mathematical Sciences (PIMS) postdoc fellowship. J. Q. is partially supported by the National

Science and Engineering Research Council of Canada (NSERC) and by the start-up funds from the University of Calgary.

References

1. J. Kennedy and R. Eberhart. Particle swarm optimization. In *Proceedings of ICNN'95-international conference on neural networks*, vol. 4, pp. 1942–1948, 1995.
2. J. Kennedy. The particle swarm: social adaptation of knowledge. In *Proceedings of 1997 IEEE International Conference on Evolutionary Computation (ICEC'97)*, pp. 303–308, 1997.
3. J. Kennedy, *Particle Swarm Optimization*, In *Encyclopedia of Machine Learning*. Springer US, Boston, MA, 2010.
4. R. Poli, J. Kennedy, and T. Blackwell, Particle swarm optimization, *Swarm intelligence*. **1** (2007), 33–57.
5. Y. Shi and R. Eberhart. A modified particle swarm optimizer. In *International conference on evolutionary computation proceedings. IEEE world congress on computational intelligence*, pp. 69–73, 1998.
6. H. M. Emara and H. A. A. Fattah. Continuous swarm optimization technique with stability analysis. In *Proceedings of the 2004 American control conference*, vol. 3, pp. 2811–2817, 2004.
7. D. Wang, D. Tan, and L. Liu, Particle swarm optimization algorithm: an overview, *Soft Comput.* (2017).
8. I. Aoki, A simulation study on the schooling mechanism in fish, *Bull. Jpn. Soc. Sci. Fish.* **48** (1982), 1081–1088.
9. A. Okubo, Dynamical aspects of animal grouping: swarms, schools, flocks, and herds, *Adv. Biophys.* **22** (1986), 1–94.
10. T. Vicsek, A. Czirók, E. Ben-Jacob, I. Cohen, and O. Shochet, Novel type of phase transition in a system of self-driven particles, *Physical Review Letters.* **75** (1995), 1226–1229.
11. I. Giardina, Collective behavior in animal groups: Theoretical models and empirical studies, *HFSP Journal.* **2** (2008), 205–219.
12. C. Reynolds. Flocks, herds and schools: A distributed behavioral model. In *Proceedings of the 14th Annual Conference on Computer Graphics and Interactive Techniques (SIGGRAPH '87). Association for Computing Machinery*, pp. 25–34, 1987.
13. J. A. Carrillo, M. Fornasier, G. Toscani, and F. Vecil. Particle, kinetic, and hydrodynamic models of swarming. In *Mathematical modeling of collective behavior in socio-economic and life sciences*, pp. 297–336. Springer, 2010.
14. D. Sumpter, *Collective animal behavior*. Princeton University Press, 2010.
15. E. Bonabeau, M. Dorigo, and G. Theraulaz, *Swarm Intelligence: From Natural to Artificial System*. Oxford University Press, 1999.
16. F. Cucker and S. Smale, Emergent behavior in flocks, *IEEE Trans. Automat. Control.* **52** (2007), 852–862.
17. S. Motsch and E. Tadmor, Heterophilious dynamics enhances consensus, *SIAM Rev.* **56** (2014), 577–621.

18. Y.-P. Choi and S. Salem, Cucker-Smale flocking particles with multiplicative noises: Stochastic mean-field limit and phase transition, *Kinetic & Related Models*. **12** (2019), 573–592.

19. T. Bäck, D. B. Fogel, and Z. Michalewicz, *Handbook of Evolutionary Computation*. IOP Publishing Ltd., 1997.

20. C. Blum and A. Roli, Metaheuristics in combinatorial optimization: Overview and conceptual comparison, *ACM computing surveys (CSUR)*. **35** (2003), 268–308.

21. M. Gendreau, J. Y. Potvin *et al.*, *Handbook of metaheuristics*. vol. 2, Springer, 2010.

22. W. Wong and C. I. Ming. A review on metaheuristic algorithms: Recent trends, benchmarking and applications. In *2019 7th International Conference on Smart Computing Communications (ICSCC)*, pp. 1–5, 2019. doi: 10.1109/ICSCC.2019.8843624.

23. J. Larson, M. Menickelly and S. M. Wild, Derivative-free optimization methods, *Acta Numerica*. **28** (2019), 287–404.

24. A. Neumaier, Complete search in continuous global optimization and constraint satisfaction, *Acta Numerica*. **13** (2004), 271–369.

25. C. Audet and W. L. Hare, *Derivative-Free and Blackbox Optimization*. Springer, 2017.

26. R. Holley and D. Stroock, Simulated annealing via Sobolev inequalities, *Communications in Mathematical Physics*. **115** (1988), 553–569.

27. S. Kirkpatrick, C. D. Gelatt and M. P. Vecchi, Optimization by simulated annealing, *Science*. **220** (1983), 671–680.

28. E. Aarts and J. Korst, *Simulated Annealing and Boltzmann Machines: A Stochastic Approach to Combinatorial Optimization and Neural Computing*. John Wiley & Sons, Inc., 1989.

29. M. Dorigo, V. Maniezzo and A. Colorni, Ant system: Optimization by a colony of cooperating agents, *IEEE Transactions on Systems, Man, and Cybernetics - Part B*. **26** (1996), 29–41.

30. M. Dorigo and C. Blum, Ant colony optimization theory: A survey, *Theoretical computer science*. **344** (2005), 243–278.

31. J. H. Holland *et al.*, *Adaptation in natural and artificial systems: an introductory analysis with applications to biology, control, and artificial intelligence*. MIT press, 1992.

32. D. Goldberg, *Genetic Algorithms in Search, Optimization and Machine Learning*. Reading, MA: Addison-Wesley Professional, 1989.

33. D. B. Fogel, *Evolutionary computation: toward a new philosophy of machine intelligence*. vol. 1, John Wiley & Sons, 2006.

34. R. Storn and K. Price, Differential evolution - a simple and efficient heuristic for global optimization over continuous spaces, *Journal of Global Optimization*. **11** (1997), 341–359.

35. K. Sörensen, Metaheuristics — the metaphor exposed, *International Transactions in Operational Research*. **22** (2015), 3–18.

36. V. Bruned, A. Mas and S. Wlodarczyk, Weak convergence of particle swarm optimization, *arXiv preprint arXiv:1811.04924*. (2019).

37. B. I. Schmitt, *Convergence Analysis for Particle Swarm Optimization.* FAU University Press, 2015.
38. G. Xu and G. Yu, On convergence analysis of particle swarm optimization algorithm, *Journal of Computational and Applied Mathematics.* **340** (2018), 709–717.
39. N. Bellomo and S. Y. Ha, A quest toward a mathematical theory of the dynamics of swarms, *Mathematical Models and Methods in Applied Sciences.* **27** (2017), 745–770.
40. R. Poli, Mean and variance of the sampling distribution of particle swarm optimizers during stagnation, *IEEE Transactions on Evolutionary Computation.* **13** (2009), 712–721.
41. Y. Zhang, S. Wang and G. Ji, A comprehensive survey on particle swarm optimization algorithm and its applications, *Mathematical problems in engineering.* **2015** (2015).
42. P.-E. Jabin and Z. Wang, Mean field limit for stochastic particle systems. In *Active Particles, Volume 1*, pp. 379–402. Springer, 2017.
43. A. S. Sznitman, Topics in propagation of chaos. In *Ecole d'été de probabilités de Saint-Flour XIX —1989*, pp. 165–251. Springer, 1991.
44. F. Golse, The mean-field limit for the dynamics of large particle systems, *Journées équations aux dérivées partielles.* (2003), 1–47.
45. P.-E. Jabin, A review of the mean field limits for Vlasov equations, *Kinetic & Related Models.* **7** (2014), 661.
46. J. A. Carrillo and Y.-P. Choi, Mean-field limits: From particle descriptions to macroscopic equations, *Arch. Ration. Mech. Anal.* **241** (2021), 1529–1573.
47. J. A. Carrillo, Y.-P. Choi and S. Salem, Propagation of chaos for the Vlasov-Poisson-Fokker-Planck equation with a polynomial cut-off, *Commun. Contemp. Math.* **21** (2019), 1850039, 28.
48. H. Huang, J.-G. Liu and P. Pickl, On the mean-field limit for the Vlasov-Poisson-Fokker-Planck system, *J. Stat. Phys.* **181** (2020), 1915–1965.
49. F. Bolley, J. A. Canizo and J. A. Carrillo, Stochastic mean-field limit: non-Lipschitz forces and swarming, *Mathematical Models and Methods in Applied Sciences.* **21** (2011), 2179–2210.
50. R. Pinnau, C. Totzeck, O. Tse and S. Martin, A consensus-based model for global optimization and its mean-field limit, *Mathematical Models and Methods in Applied Sciences.* **27** (2017), 183–204.
51. J. A. Carrillo, Y.-P. Choi, C. Totzeck and O. Tse, An analytical framework for consensus-based global optimization method, *Mathematical Models and Methods in Applied Sciences.* **28** (2018), 1037–1066.
52. J. A. Carrillo, S. Jin, L. Li and Y. Zhu, A consensus-based global optimization method for high dimensional machine learning problems. In *ESAIM: Control, Optimisation and Calculus of Variations*, 2020.
53. C. Totzeck and M. T. Wolfram, Consensus-based global optimization with personal best, *Mathematical Biosciences and Engineering.* **17** (2020), 6026–6044.
54. C. Totzeck, R. Pinnau, S. Blauth and S. Schotthöfer, A numerical comparison of consensus-based global optimization to other particle-based global optimization schemes, *PAMM.* **18** (2018), 1–28.

55. S. Y. Ha, S. Jin and D. Kim, Convergence of a first-order consensus-based global optimization algorithm, *Mathematical Models and Methods in Applied Sciences.* **30** (2020), 2417–2444.

56. S. Y. Ha, S. Jin and D. Kim, Convergence and error estimates for time-discrete consensus-based optimization algorithms, *Numerische Mathematik.* **147** (2021), 255–282.

57. J. Chen, S. Jin and L. Lyu, A consensus-based global optimization method with adaptive momentum estimation, *arXiv preprint arXiv:2012.04827.* (2020).

58. M. Fornasier, T. Klock and K. Riedl, Consensus-based optimization methods converge globally in mean-field law, *arXiv preprint arXiv:2103.15130.* (2021).

59. C. Totzeck, Trends in consensus-based optimization, In: Bellomo, N., Carrillo, J.A., Tadmor, E. (eds.), Active Particles, Volume 3. *Modeling and Simulation in Science, Engineering and Technology.* Birkhäuser, Cham (2022), 201–226.

60. M. Fornasier, H. Huang, L. Pareschi and P. Sünnen, Consensus-based optimization on hypersurfaces: Well-posedness and mean-field limit, *Mathematical Models and Methods in Applied Sciences.* **30** (2020), 2725–2751.

61. M. Fornasier, H. Huang, L. Pareschi and P. Sünnen, Consensus-based optimization on the sphere: Convergence to global minimizers and machine learning, *J. Machine Learning Research.* **22** (2021), 1–55.

62. M. Fornasier, H. Huang, L. Pareschi and P. Sünnen, Anisotropic diffusion in consensus-based optimization on the sphere, *SIAM J. Optim.* **32** (2022), 1984–2012.

63. A. Benfenati, G. Borghi and L. Pareschi, Binary interaction methods for high dimensional global optimization and machine learning, *Applied Math. Optim.* **86** (2022), 1–41.

64. L. Pareschi and G. Toscani, *Interacting multiagent systems: kinetic equations and Monte Carlo methods.* OUP Oxford, 2013.

65. G. Naldi, L. Pareschi and G. Toscani, eds., *Mathematical Modeling of Collective Behavior in Socio-Economic and Life Sciences*, vol. 10, *Modeling and Simulation in Science, Engineering and Technology.* Birkhäuser Basel, 2010.

66. J. A. Carrillo, F. Hoffmann, A. M. Stuart and U. Vaes, Consensus based sampling, *Studies in Applied Mathematics* **148** (2022), 1069–1140.

67. G. Borghi, S. Grassi and L. Pareschi, *Consensus based optimization with memory effects: fast algorithms and implementation aspects*, preprint (2022).

68. S. Grassi and L. Pareschi, From particle swarm optimization to consensus based optimization: stochastic modeling and mean-field limit, *Mathematical Models and Methods in Applied Sciences.* **31** (2021), 1625–1657.

69. H. Huang, A note on the mean-field limit for the particle swarm optimization, *Applied Mathematics Letters.* **117** (2021), 107133.

70. H. Huang and J. Qiu, *Mathematical Methods in the Applied Sciences*, **45** (2022), 7814–7831.

71. C. Cipriani, H. Huang and J. Qiu, Zero-inertia limit: from particle swarm optimization to consensus based optimization, *SIAM Journal on Mathematical Analysis*, **54** (2022), 3091–3121.

72. H. Huang, J. Qiu and K. Riedl, On the global convergence of particle swarm optimization methods, arXiv preprint arXiv:2201.12460 (2021).

73. J.-A. Acebrón and R. Spigler, Adaptive frequency model for phase-frequency synchronization in large populations of globally coupled nonlinear oscillators, *Physical Review Letters.* **81** (1998), 2229–2232.

74. M. H. Duong, A. Lamacz, M. A. Peletier and U. Sharma, Variational approach to coarse-graining of generalized gradient flows, *Calculus of variations and partial differential equations.* **56** (2017), 65–100.

75. J. H. M. Evers, R. C. Fetecau and W. Sun, Small inertia regularization of an anisotropic aggregation model, *Math. Models Methods Appl. Sci.* **27** (2017), 1795–1842.

76. R. C. Fetecau and W. Sun, First-order aggregation models and zero inertia limits, *J. Diff. Equations.* **259** (2015), 6774–6802.

77. Y.-P. Choi and O. Tse, Quantified overdamped limit for kinetic Vlasov-Fokker-Planck equations with singular interaction forces, *Journal of Differential Equations*, **330** (2022), 150–207.

78. J. A. Carrillo, Y.-P. Choi and Y. Peng, Large friction-high force fields limit for the nonlinear Vlasov-Poisson-Fokker-Planck system, *Kinetic and Related Models*, **15** (2022), 355–384.

79. G. Albi and L. Pareschi, Binary interaction algorithms for the simulation of flocking and swarming dynamics, *Multiscale Modeling & Simulation.* **11** (2013), 1–29.

80. S. Jin, L. Li and J. G. Liu, Random Batch Methods (RBM) for interacting particle systems, *Journal of Computational Physics.* **400** (2020), 108877.

81. C. M. Bishop, *Pattern recognition and machine learning.* Springer, 2006.

82. V. Vapnik, Principles of risk minimization for learning theory. In *Advances in neural information processing systems*, pp. 831–838, 1991.

83. P. Jain and P. Kar, Non-convex optimization for machine learning, *Foundations and Trends in Machine Learning.* **10** (2017), 142–363.

84. R. Durrett, *Stochastic calculus: a practical introduction.* CRC press, 2018.

85. P. Billingsley, *Convergence of probability measures.* John Wiley & Sons, 2013.

86. H. P. McKean Jr., A class of Markov processes associated with nonlinear parabolic equations, *Proceedings of the National Academy of Sciences of the United States of America.* **56** (1966), 1907.

87. G. Da Prato and J. Zabczyk, *Stochastic equations in infinite dimensions.* Cambridge university press, 2014.

88. X. Mao, *Stochastic differential equations and applications.* Elsevier, 2007.

89. E. Platen, An introduction to numerical methods for stochastic differential equations, *Acta numerica.* **8** (1999), 197–246.

90. G. Dimarco and L. Pareschi, Numerical methods for kinetic equations, *Acta Numerica.* **23** (2014), 369–520.

91. M. Jamil and X. S. Yang, A literature survey of benchmark functions for global optimisation problems, *International Journal of Mathematical Modelling and Numerical Optimisation.* **4** (2013), 150–194.

Consensus-based Optimization and Ensemble Kalman Inversion for Global Optimization Problems with Constraints

José Antonio Carrillo

Mathematical Institute
University of Oxford
Oxford OX2 6GG, UK
carrillo@maths.ox.ac.uk

Claudia Totzeck

University of Wuppertal
School of Mathematics and Natural Sciences
Gaußstr. 20, 42119 Wuppertal, Germany
totzeck@uni-wuppertal.de

Urbain Vaes

MATHERIALS team
Inria Paris
Paris 75012, France
urbain.vaes@inria.fr

We introduce a practical method for incorporating equality and inequality constraints in global optimization methods based on stochastic interacting particle systems, specifically consensus-based optimization (CBO) and ensemble Kalman inversion (EKI). Unlike other approaches in the literature, the method we propose does not constrain the dynamics to the feasible region of the state space at all times; the particles evolve in the full space, but are attracted towards the feasible set by means of a penalization term added to the objective function and, in the case of CBO, an additional relaxation drift. We study the properties of the method through the associated mean-field Fokker–Planck equation and demonstrate its performance in numerical experiments on several test problems.

1. Introduction

Finding efficiently the global minimizer of a function in large dimensions is crucial in today's data science activity, with applications in various fields of research ranging from modern areas such as machine learning, neural networks, inverse problems, data assimilation and model parameter estimation to more classical areas such as economics, computational biology, mathematical finance and statistical physics. In the most general form, the global optimization task is simply

$$\underset{x \in B}{\operatorname{argmin}} f(x) \qquad (1.1)$$

for a given objective function f and state space $B \subset \mathbf{R}^d$. This simple-to-describe problem is highly nontrivial for general nonconvex functions f due to the existence of usually many, and in some cases even infinitely many, local minima. Most of the approaches have followed metaheuristic arguments in order to find robust algorithms; see for instance Refs. [1–6] and the references therein. Among global optimization methods, several algorithms were inspired by physical or biological models using the idea of interacting particle/agent systems: ant colony optimization [7], artificial bee colony optimization [8], firefly optimization [9], wind driven optimization [10], and particle swarm optimization [11, 12]. All of these methods have in common the use of interactions of particles with nonlocal terms, the exchange of information among agents and the use of noise terms to explore the landscape of the function f; see Ref. [13] for a review. Another family of global optimization algorithms is known as simulated annealing methods [14,15]. The simplest examples of simulated annealing methods use the cost function f as the potential in an overdamped Langevin diffusion, which is solved with a time-dependent temperature [16–18] — a cooling schedule — so that the unique probability measure in the kernel of the Fokker–Planck operator associated with the dynamics at time t converges to a Dirac mass at the global minimizer in the limit as $t \to \infty$. Despite their simplicity, these methods usually have a slow convergence rate.

Concerning machine learning applications and large scale optimization, there have been a flurry of works in SGD methods, where stochasticity is again introduced as a way of improving exploration of the landscape. An essential ingredient of SGD methods [19–21] is the mini-batch strategy, also introduced recently in Refs. [22, 23] for interacting particle systems, which reduces the computational cost of the iterations. The SGD family of methods relies on the computation of gradients of functions in high-dimensional settings, which may be computationally expensive or even impossible, if the loss function is nondifferentiable. In addition, even when gradients can be calculated at a reasonable computational cost, they may fail to provide useful information on the overall behavior of the loss function if this function exhibits many small-scale oscillations. Gradient-free methods, which rely only on evaluations of the loss function, are therefore an attractive alternative to SGD methods in these settings. In this chapter, we consider two particular classes of methods belonging to this category which received a lot of attention lately: the Consensus-Based Optimization (CBO) methods [24–27], reviewed recently in Ref. [28], and methods based on the Ensemble Kalman Filter (EnKF) [29–35], which are mainly employed in the context of inverse problems but have also proved useful for machine learning tasks [36]. It is shown in Ref. [37] that gradient-free ensemble Kalman methods perform better than their gradient-based counterparts in noisy likelihood landscapes. Within the class of ensemble Kalman methods, we focus specifically on the Ensemble Kalman Inversion (EKI) gradient-free method, in the form proposed in Ref. [32].

While the EKI method as discussed in Sec. 3 relies on a nonlocal approximation of gradients, and can be viewed as an exact preconditioned gradient descent in the case of a quadratic objective function [33], the central mechanism of CBO is a relaxation drift towards the weighted average

$$
m_f = \frac{\sum_{j=1}^{J} x^{(j)} \mathrm{e}^{-\alpha f(x^{(j)})}}{\sum_{j=1}^{J} \mathrm{e}^{-\alpha f(x^{(j)})}} =: \sum_{j=1}^{J} w(x^{(j)}) x^{(j)}. \tag{1.2}
$$

Here, $x^{(j)}$ are the positions of a number J of explorers evolving in the landscape of the objective function f. As a metaphor for the CBO method, one may think of butterflies in the mountains able to communicate their height and position instantaneously by wireless communication and to integrate this information in the form of the weighted average m_f, towards which they move with the hope of eventually reaching the global minimum of the landscape given by the function f.

The motivation for the CBO reweighting $\mathrm{e}^{-\alpha f(x)}$, which is the Gibbs distribution corresponding to $f(x)$, comes from statistical mechanics: the probability distribution function $\frac{1}{Z_\alpha} \mathrm{e}^{-\alpha f(x)}$, where Z_α is the normalization constant, is the unique invariant measure of overdamped Langevin dynamics at temperature α^{-1} in an external potential given by the cost function $f(x)$ [38, Chapter 6]. For sufficiently large α, the weighted average (1.2) provides a good approximation of the particle position $x^{(j_*)}$ minimizing $f(x^{(j)})$, assuming this is unique. Indeed, in this limit all the weights in (1.2) converge to 0, except for the one associated with particle j_*, which converges to 1.

Both the CBO and EKI methods are based on a system of interacting stochastic differential equations, and it is possible to show that their behavior is well-described by nonlocal deterministic Fokker–Planck equations when the number of particles is sufficiently large. Studying these continuous equations brings considerable insight, and turns out to be much simpler than working with the particle systems. The mean-field Fokker–Planck equation associated with CBO includes a weighted average which, despite no longer being a finite sum as in (1.2), can still be understood in the limit as $\alpha \to \infty$. This is achieved through Laplace's method [39, 40], a classical asymptotic method for integrals.

In its original form [24], the CBO method combines the relaxation drift towards the weighted average m_f with isotropic multiplicative noise proportional in amplitude to the Euclidean distance to the weighted average m_f. The specific form of the multiplicative factor of the noise, which may be viewed as a temperature scheduling, is improved in Ref. [27] in order to decrease the computational cost of the method for large dimensional problems; see Sec. 2 for details. In Ref. [27], the authors also demonstrate that approximating m_f based on a randomly selected subset of the particles — a mini-batch — leads to large computational savings. Using mini-batches also introduces extra stochasticity, which is observed to be useful for promoting exploration of the landscape given by f.

In this chapter, we propose new extensions of the CBO and EKI methods to constrained optimization and constrained inverse problems, respectively. In the case of CBO, we consider problems of the general form

$$\operatorname*{argmin}_{x \in B} f(x), \quad \text{with} \quad B = \left\{ x : \mathcal{E}(x) = 0, \mathcal{I}(x) \geq 0 \right\}, \qquad (1.3)$$

where $\mathcal{E} \colon \mathbf{R}^d \to \mathbf{R}^{N_e}$ and $\mathcal{I} \colon \mathbf{R}^d \to \mathbf{R}^{N_i}$ are continuously differentiable functions and the inequality $\mathcal{I}(x) \geq 0$ is understood componentwise. In the case of EKI, we consider subsets of problems of the form (1.3) for which the objective function is of the specific form $f(x) = \frac{1}{2} \left| \Gamma^{-1} \big(G(x) - y \big) \right|^2$, where $G : \mathbf{R}^d \to \mathbf{R}^K$ is a map, $\Gamma \in \mathbf{R}^{K \times K}$ is a positive definite matrix, and $y \in \mathbf{R}^K$. In applications, objective functions of this form are used as a simple measure of the misfit between some data y and a model $G(x)$.

The literature on global optimization with constraints is abundant and an extensive review is beyond the scope of this chapter, so we summarize hereafter only recent contributions that are specifically related to the CBO and EKI methods. Several recent works propose extensions of CBO strategies to equality constrained problems, based on imposing that the dynamics is restricted to the feasible manifold at all times [41–43]. Most of these works consider the specific case of the Euclidean sphere as the feasible region, with applications to robust subspace detection and efficient eigenvalue computations in machine learning as prime examples. The evolution of particles is restricted to the feasible manifold by appropriate projection onto the tangent space. Although this method performs relatively well in very simple settings like the sphere, its implementation is difficult and error-prone for more general constraint manifolds. In addition, since the weighted average is computed in the ambient space, there are cases in which the method does not produce good results, for example when the feasible region is a closed hypersurface enclosing a nonconvex domain. Let us finally emphasize that manifolds other than the sphere are important in optimization problems in machine learning. SGD methods have been studied in Riemannian optimization applications, see for instance Refs. [44, 45] and the references therein. Here, the state space is usually a set of matrices with certain constraints, like positive definite matrices in Wasserstein barycenters or Riemannian centroids.

Several methods have also been proposed in the literature for incorporating constraints in EKI. In Ref. [46], the authors propose a generalization of EKI such that the iterates produced by the method are guaranteed to lie in the feasible region at each iteration. The method proposed leverages the fact that the update step in the usual ensemble Kalman method can be formulated as an optimization problem, in which linear constraints can be integrated. In Ref. [47], a variant of EKI incorporating a projection step is developed for the specific case of box constraints, and the continuous time limit of the method is studied. See also Ref. [48], where the continuous-time and mean-field limits of the method proposed in Ref. [46] are studied.

In this chapter, we follow an orthogonal approach: we propose extensions of the CBO and EKI methods in which the particles evolve in the full space $\mathbf{R}^d$, instead of being constrained to the feasible manifold at all times. A requirement for this approach to work, which we take as a standing assumption in the rest of this chapter, is that the objective function $f(x)$ can be evaluated anywhere in $\mathbf{R}^d$, which may or may not hold in applications. The unifying idea behind our extensions of CBO and EKI is the addition of a penalization to the objective function. In the case of only one equality constraint $\mathcal{E}(x) = 0$, for example, we seek a solution to

$$\underset{x \in \mathbf{R}^d}{\operatorname{argmin}} \left(f(x) + \frac{1}{\nu} |\mathcal{E}(x)|^2 \right). \tag{1.4}$$

When the penalization parameter ν tends to 0, we expect the solution of this problem to be close to the zero-level set of $\mathcal{E}$. We emphasize that the use of penalty functions for constrained optimization problems is a standard idea and can be employed, in principle, together with any method for unconstrained global optimization. In this chapter, we demonstrate the effectiveness of this simple idea when used in conjunction with CBO and EKI, and we propose specific approaches for its implementation.

In the specific case of CBO, we also propose to use, in addition to a penalization of the form (1.3), an extra drift term based on the constraint that imposes that particles are asymptotically confined to the manifold. This idea is reminiscent of swarming problems where the Vicsek model with noise on the sphere can be retrieved from the Cucker-Smale model on the whole space in an appropriate limit [49].

The rest of this chapter is organized as follows. In Secs. 2 and 3, we explain our strategy for including constraints in the CBO and the EKI methods, respectively. In Sec. 2, we also compare the solution to the constrained CBO method and associated the mean-field equations qualitatively for a toy problem in two dimensions, in order to illustrate the method and gain a deeper understanding of its behavior in the many-particle limit. In Secs. 4 and 5, we investigate the potential of the proposed methods by means of numerical experiments in some reference examples. To this end, we use typical benchmarks [50] in optimization. Section 6 is reserved for conclusions and perspectives for future work.

2. CBO for constrained global optimization problems

In order to illustrate our strategy for incorporating constraints into the CBO methods and its variants, we consider the CBO scheme with component-wise Brownian motion proposed in Ref. [26], which reads as

$$\mathrm{d}x_t^{(j)} = -(x_t^{(j)} - m_f)\,\mathrm{d}t + \sqrt{2}\sigma(x_t^{(j)} - m_f) \circ \mathrm{d}W_j^{(j)}, \quad j = 1, \ldots, J, \tag{2.1}$$

where $\{W^{(j)}\}_{1 \le j \le J}$ are independent Wiener processes in $\mathbf{R}^d$ and $v_1 \circ v_2$, for vectors v_1 and v_2 in $\mathbf{R}^d$, is the Hadamart (component-wise) product, that is to say $v_1 \circ v_2 = \operatorname{diag}(v_1)v_2$. The particles are initialized independently according to some given probability density ϱ_0, and so the law of the initial ensemble is $\varrho_0^{\otimes J}$. The main ingredient of the method, indeed the only mechanism by which the particles interact with the objective function, is the weighted mean m_f given by

$$m_f(\mu_t^J) = \frac{\int x e^{-\alpha f(x)} \mu_t^J(\mathrm{d}x)}{\int e^{-\alpha f(x)} \mu_t^J(\mathrm{d}x)}, \qquad \mu_t^J = \frac{1}{J} \sum_{j=1}^{J} \delta_{x_t^{(j)}}, \qquad (2.2)$$

where $\alpha > 0$ is a parameter. Equivalently,

$$m_f(\mu_t^J) = \frac{\sum_{j=1}^{J} x_t^{(j)} \exp\left(-\alpha f(x_t^{(j)})\right)}{\sum_{j=1}^{J} \exp\left(-\alpha f(x_t^{(j)})\right)}.$$

As mentioned in the introduction, for fixed t the weighted mean $m_f(\mu_t^J)$ in (2.2) converges to the global minimizer of f constrained to the support of the empirical measure μ_t^J, provided this minimizer is unique, in the limit as α tends to infinity.

Assume the problem we aim to solve is (1.1). In order to illustrate the incorporation of equality and inequality constraints in a unified manner, we introduce $\mathcal{A} \colon \mathbf{R}^d \to \mathbf{R}^{N_e + N_i}$ given by

$$(\mathcal{A}(x))_i = \begin{cases} (\mathcal{E}(x))_i & \text{if } i \le N_e, \\ \min\{(\mathcal{I}(x))_{i - N_e}, 0\} & \text{if } i > N_e. \end{cases}$$

We can then write $B = \{x \in \mathbf{R}^d : \mathcal{A}(x) = 0\}$; that is, we can assume without loss of generality that all the constraints are of equality type. Notice that $\mathcal{A}$ may not be C^1 at the manifold even though $\mathcal{E}$ and $\mathcal{I}$ are, but this is not an issue given that only $\nabla |\mathcal{A}|^2$ appears in the method we propose, and the function $x \mapsto |\mathcal{A}(x)|^2$ is continuously differentiable. To solve the optimization problem (1.1) with constraint $\mathcal{A}(x) = 0$, we modify the scheme (2.1) in the following manner:

(1) The first modification is a penalization of the objective function in the weighted average. More precisely, we substitute m_f in (2.1) by

$$m_g(\mu_t^J) = \frac{\int x e^{-\alpha g(x)} \mu_t^J(\mathrm{d}x)}{\int e^{-\alpha g(x)} \mu_t^J(\mathrm{d}x)}, \qquad g(x) := f(x) + \frac{1}{\nu} |\mathcal{A}(x)|^2, \qquad (2.3)$$

where ν is a small parameter. By the Laplace principle, for fixed t this new weighted average tends to the minimizer of g (within the support of the empirical measure and assuming this minimizer is unique) in the limit as $\alpha \to \infty$, which is expected to lie close to the feasible set $\{x : \mathcal{A}(x) = 0\}$ when the parameter ν is sufficiently small.

(2) The second modification is the introduction of a drift term that drives particles towards the constraint manifold. For $0 < \varepsilon \ll 1$ we propose the dynamics

$$
\begin{aligned}
\mathrm{d}x_t^{(j)} = {}&-\frac{1}{\varepsilon}(\nabla|\mathcal{A}|^2)(x_t^{(j)})\,\mathrm{d}t - \left(x_t^{(j)} - m_g(\mu_t^J)\right)\mathrm{d}t \\
&+ \sqrt{2}\sigma\left(x_t^{(j)} - m_g(\mu_t^J)\right)\circ \mathrm{d}W_t^{(j)}.
\end{aligned} \tag{2.4}
$$

Both mechanisms are useful for driving the particles to the constraint manifold. The first modification is straightforward to implement and does not generate stiffness of the stochastic differential equations driving the particles. Consequently, these can be discretized in time with a step of the same order of magnitude as that used for unconstrained problems. The second modification is helpful for ensuring that the particles move towards the weighted average in a manner independent of the other particles. The additional drift pushes the particles to the manifold in an asymptotically orthogonal manner (indeed $\nabla|\mathcal{A}|^2$ is orthogonal to the contour lines of $|\mathcal{A}|^2$), giving us more function evaluations along the manifold. Despite being detrimental for stability at the discrete level, this additional drift is observed to be useful in our numerical experiments.

Example 2.1: For global optimization problems subject to just one inequality constraint, i.e. problems of the form

$$
\operatorname*{argmin}_{x\in B} f(x) \quad\text{with}\quad B = \{x : \mathcal{I}(x) \geq 0\}
$$

with $\mathcal{I}\colon \mathbf{R}^d \to \mathbf{R}$, the above scheme reads

$$
\begin{aligned}
\mathrm{d}x_t^{(j)} = {}&-\frac{1}{\varepsilon}\nabla(\mathcal{I}^2)(x_t^{(j)})\chi_{(-\infty,0)}\big(\mathcal{I}(x_t^{(j)})\big)\,\mathrm{d}t - (x_t^{(j)} - m_g(\mu_t^J))\,\mathrm{d}t \\
&+ \sqrt{2}\sigma(x_t^{(j)} - m_g(\mu_t^J))\circ \mathrm{d}W_t^{(j)},
\end{aligned}
$$

where $\chi_{(-\infty,0)}$ denotes the indicator function of the set $(-\infty,0)$. The weighted average is given by

$$
m_g(\mu_t^J) = \frac{\displaystyle\sum_{j=1}^{J} x_t^{(j)}\exp\left(-\alpha f(x_t^{(j)}) - \alpha\nu^{-1}\mathcal{I}(x_t^{(j)})^2\chi_{(-\infty,0)}\big(\mathcal{I}(x_t^{(j)})\big)\right)}{\displaystyle\sum_{j=1}^{J}\exp\left(-\alpha f(x_t^{(j)}) - \alpha\nu^{-1}\mathcal{I}(x_t^{(j)})^2\chi_{(-\infty,0)}\big(\mathcal{I}(x_t^{(j)})\big)\right)}.
$$

2.1. *Mean-field CBO*

It is possible to prove a propagation of chaos result for (2.1); see Ref. [51]. In the many-particle limit $J \gg 1$, the law of $(x_t^{(1)}, \ldots, x_t^{(J)})$ approximately tensorises, i.e. it holds approximately that

$$(x_t^{(1)}, \ldots, x_t^{(J)}) \sim \mu_t^{\otimes J}, \tag{2.5}$$

where μ_t is a probability measure. In addition, this measure satisfies the partial differential equation

$$\partial_t \mu = \nabla \cdot \left((x - m_f(\mu)) \mu \right) + \sigma^2 \sum_{i=1}^{d} \partial_{ii} \left((x - m_f(\mu))_i^2 \mu \right), \quad \mu_0 = \varrho_0, \tag{2.6}$$

in the distributional sense. In order to formally derive this equation from (2.5), one can take the limit $J \to \infty$ in (2.1) and use the law of large numbers to obtain

$$dx_t^{(1)} = -\left(x_t^{(1)} - m_f(\mu_t) \right) dt + \sqrt{2}\sigma \left(x_t^{(1)} - m_f(\mu_t) \right) \circ dW_t^{(1)}. \tag{2.7}$$

The Fokker–Planck equation corresponding to this equation is then (2.6). A similar formal argument for (2.4) leads to the following non-local Fokker–Planck equation

$$\partial_t \mu = \frac{1}{\varepsilon} \nabla \cdot (\nabla |\mathcal{A}|^2(x)\mu) + \mathrm{div}_x \left((x - m_g(\mu))\mu \right)$$
$$+ \sigma^2 \sum_{i=1}^{d} \partial_{ii} \left((x - m_g(\mu))_i^2 \mu \right), \quad \mu_0 = \varrho_0. \tag{2.8}$$

Remark 2.2: If g in (2.3) satisfies the assumptions in Ref. [25], then all the theoretical results proved there for the mean-field Fokker–Planck equation associated with unconstrained CBO apply mutatis mutandis to (2.6) when $\varepsilon = \infty$, i.e. with only the first modification — the penalization.

2.2. *Numerical experiments: mean-field versus particles and the penalizations*

We illustrate the CBO approach with and without constraints by comparing the solution to the mean-field PDE (2.6) with the solution to the particle system (2.1) in simple two dimensional problems.

We begin with a comparison in the unconstrained setting. To this end, we use the Ackley benchmark [50] shifted by $(0.5, 0)$; see Fig. 1(left) for a contour plot. The SDE simulation uses 50 particles, which are drawn independently from the normal distribution $\mathcal{N}(0, 3I_2)$, where I_2 is the 2 by

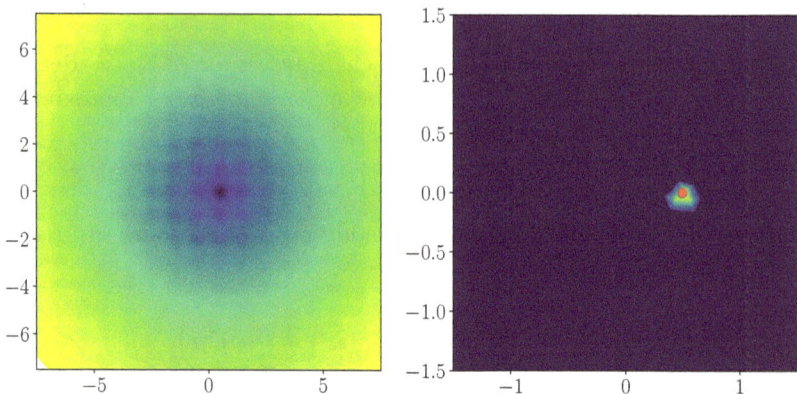

Fig. 1. $J = 50$, $M = 100$, $\Delta t_{\mathrm{SDE}} = 0.0005$, $\Delta t_{\mathrm{PDE}} = 0.005$, $h_{\min} = 0.125$, $h_{\max} = 0.5$, $\alpha = 30$, $\nu = 1$, $\sigma = 0.7$, $T = 50$. Here $h_{\min}$ and $h_{\max}$ denote the minimum and maximum characteristic lengths of the mesh employed for the PDE simulation. Left: contour plot of the Ackley function shifted by $(0.5, 0)$. Right: contour plot of the PDE solution which is concentrated at one triangle. The red points show the weighted averages of 100 samples of particle simulations after convergence.

2 identity matrix, at the beginning of the simulation. The PDE solution is initialized according to the same distribution. In view of the shift, the global minimizer of the objective function is not at the center of mass of the initial distribution. The mesh for the PDE solver has a maximal meshfield size of $h_{\max} = 0.5$ and is refined around the origin with a minimal meshfield size of $h_{\min} = 0.125$. In particular, the region around the global minimum is contained in the refined area. Additional details on the numerical implementation and the parameters employed in the simulations are given in Sec. 4.

Figure 1(right) illustrates the weighted means after convergence of $M = 100$ independent simulations, as well as the contour plot of the PDE solution after convergence to one triangle. The weighted means lie at the minimizer of the shifted Ackley function and in particular in the support of the PDE solution. This demonstrates that the SDE and PDE models are in good agreement, even with as few as 50 particles. Note that the left panel in Fig. 1 covers the domain $[-7.5, 7.5]^2$, while the right panel covers the zoomed-in region $[-1.5, 1.5]^2$.

Next, we study the influence of the penalization (2.3). We consider again the Ackley function shifted by $(0.5, 0)$, with now a constraint given by the circle $B = \{x \in \mathbf{R}^2 : x^2 = 9\}$, which is depicted in gray in Fig. 2.

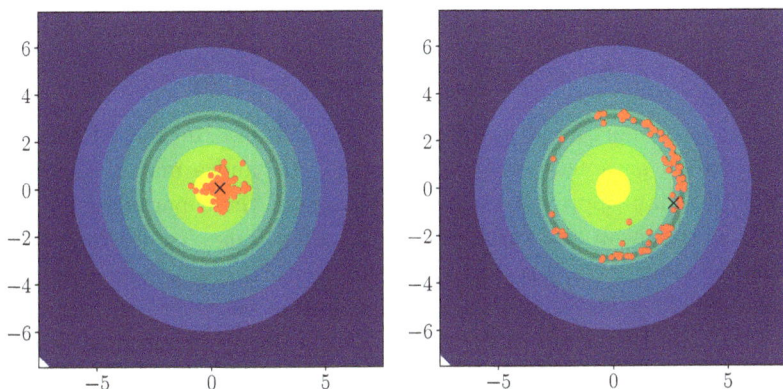

Fig. 2. $J = 50, M = 100, T = 0, \alpha = 30, \nu = 0/1$. Left: weighted averages with $\nu = 0$ of 100 samples at the beginning of the simulation. Right: weighted averages with $\nu = 1$ of 100 samples at the beginning of the simulation. The contour plots show the initial distribution of the mean-field solution. The influence of the constraint in the weight is clearly visible.

In this figure, the filled contours illustrate the initial density of the PDE solution, while the red dots are the weighted averages corresponding to $M = 100$ independent particle simulations at the initial time. The black crosses indicate the positions of the weighted averages of the PDE solution at the initial time, without (left, calculated using (1.2)) or with (right, calculated using (2.3)) penalization. We observe that the weighted average with penalization of the initial PDE solution is already very close to the true global minimizer of the constrained problem. This numerical experiment motivates the strategy of enforcing the constraints through the penalization on the objective functions.

3. EKI with constraints

The EKI was initially proposed as a method for solving inverse problems of the following form: find an unknown parameter $x \in \mathbf{R}^d$ from data $y \in \mathbf{R}^K$ given that

$$y = G(x) + \eta, \qquad \eta \sim \mathcal{N}(0, \Gamma), \tag{3.1}$$

where $G : \mathbf{R}^d \to \mathbf{R}^K$ is a nonlinear map known as the *forward operator*. A point estimator for the solution to (3.1) is usually defined as the minimizer of a least-squares functional of the form

$$\Phi_R(x; y) = \frac{1}{2}|y - G(x)|_\Gamma^2 + \frac{1}{2}|x - a|_\Sigma^2. \tag{3.2}$$

Here $|\cdot|_A$ is a short-hand notation for $|A^{-1/2}\cdot|$ and the second term is a regularization parametrized by a vector $a \in \mathbf{R}^d$ and a matrix $\Sigma \in \mathbf{R}^{d \times d}$. In the Bayesian approach to inverse problems, these parameters can be viewed as the mean and variance of a Gaussian *prior distribution* $\mathcal{N}(a, \Sigma)$ on the unknown x. In this case, the minimizer of $\Phi_R(\cdot\,; y)$ is the pointwise maximizer of the Bayesian posterior distribution associated with the problem [52]. The EKI is an optimization scheme for the functional $\Phi_R(\cdot\,; y)$ that can be derived from the ensemble Kalman filter. Its continuous-time version is based on the following interacting particle system:

$$dx_t^{(j)} = -\frac{1}{J} \sum_{k=1}^J \langle G(x_t^{(k)}) - \bar{G}_t, G(x_t^{(j)}) - y\rangle_\Gamma \left(x_t^{(k)} - \bar{x}_t\right) dt$$

$$- \mathcal{C}(\mu_t^J)\Sigma^{-1}(x_t^{(j)} - a)\,dt, \qquad j = 1, \ldots, J, \tag{3.3}$$

where $\mu_t^J = \frac{1}{J}\sum_{j=1}^J \delta_{x_t^{(j)}}$ denotes as before the empirical measure associated with the ensemble, and the quantities $\bar{x}_t$ and $\bar{G}_t$ are defined as

$$\bar{x}_t = \int_{\mathbf{R}^d} x\,\mu_t^J(dx) = \frac{1}{J}\sum_{j=1}^J x_t^{(j)},$$

$$\bar{G}_t = \int_{\mathbf{R}^d} G(x)\,\mu_t^J(dx) = \frac{1}{J}\sum_{j=1}^J G(x_t^{(j)}).$$

The matrix $\mathcal{C}(\mu_t^J)$ is the covariance under the empirical distribution:

$$\mathcal{C}(\mu_t^J) = \int_{\mathbf{R}^d} (x \otimes x)\mu_t^J(dx) = \frac{1}{J}\sum_{j=1}^J (x_t^{(j)} - \bar{x}_t) \otimes (x_t^{(j)} - \bar{x}_t).$$

The first argument in the inner product on the right-hand side of (3.3) is related to consensus, whereas the second argument measures the mismatch with the observed data. In the limit $t \to \infty$, the particles are expected to concentrate at the global minimizer of Φ_R given in (3.2).

In the case of a linear forward map G, the EKI algorithm takes the form of a gradient descent preconditioned by the ensemble covariance:

$$dx_t^{(j)} = -\mathcal{C}(\mu_t^J)\nabla\Phi_R(x_t^{(j)})\,dt, \qquad j = 1, \ldots, J. \tag{3.4}$$

In the case of a general forward map, the EKI (3.3) can be viewed as a derivative-free approximation of (3.4). This viewpoint, which is adopted in Refs. [33, 34], is useful below: we first show how to include constraints for (3.4), and then perform a derivative-free approximation of the resulting system.

In the Bayesian approach to inverse problems, it would be natural to incorporate equality and inequality constraints in the prior distribution on the unknown parameter x. The EKI method, however, applies only to the case of a Gaussian prior distribution, i.e. regularization by a quadratic function in (3.2), and so extensions of this algorithm are required in order to handle nonlinear constraints.

In this section, we demonstrate how a simple penalty-based method similar to that in Sec. 2 for CBO can be employed for including constraints in EKI. For the sake of simplicity, we assume there is only one equality constraint $\mathcal{A}(x) = 0$. This constraint can be incorporated in the gradient method (3.4) by adding a penalization term to the regularized least-squares functional:

$$
\begin{aligned}
\mathrm{d}x_t^{(j)} &= -\mathcal{C}(\mu_t^J)\nabla\left(\Phi_R + \frac{1}{\nu}|\mathcal{A}|^2\right)(x_t^{(j)})\,\mathrm{d}t, \\
&= -\mathcal{C}(\mu_t^J)\nabla\Phi_R(x_t^{(j)})\,\mathrm{d}t - \frac{2}{\nu}\mathcal{C}(\mu_t^J)\mathcal{A}(x_t^{(j)})\nabla\mathcal{A}(x_t^{(j)})\,\mathrm{d}t, \quad j = 1,\ldots,J.
\end{aligned}
$$

A derivative-free version of this methodology is obtained by approximating gradients in the same manner as in standard EKI:

$$
\begin{aligned}
\mathrm{d}x_t^{(j)} &= -\frac{1}{J}\sum_{k=1}^{J}\langle G(x_t^{(k)}) - \bar{G}_t, G(x^{(j)}) - y\rangle_\Gamma(x^{(k)} - \bar{x}_t)\mathrm{d}t \\
&\quad - \mathcal{C}(\mu_t^J)\Sigma^{-1}(x^{(j)} - a)\,\mathrm{d}t - \frac{2}{\nu}\mathcal{C}(\mu_t^J)\mathcal{A}(x_t^{(j)})\nabla\mathcal{A}(x_t^{(j)})\,\mathrm{d}t,
\end{aligned}
\tag{3.5}
$$

for $j = 1,\ldots,J$. In contrast with the other methods proposed in the literature, the particles forming the ensemble produced by this method are not confined to the feasible region. Therefore, as already emphasized in the introduction, a prerequisite of the method is that the forward map may be evaluated for any $x \in \mathbf{R}^d$, and not only in the feasible region.

The formulation (3.5) also assumes that the gradient of the function $\mathcal{A}$ can be calculated analytically. In the general case, one may employ a gradient-free approximation of $\nabla|\mathcal{A}|^2$,

$$dx_t^{(j)} = -\frac{1}{J}\sum_{k=1}^{J}\langle G(x_t^{(k)}) - \bar{G}_t, G(x_t^{(j)}) - y\rangle_\Gamma x_t^{(k)}dt - C(\mu_t^J)\Sigma^{-1}(x_t^{(j)} - a)dt,$$

$$-\frac{2}{\nu J}\sum_{k=1}^{J}\langle \mathcal{A}(x_t^{(k)}) - \bar{\mathcal{A}}_t, \mathcal{A}(x_t^{(j)})\rangle(x_t^{(k)} - \bar{x}_t)\,dt, \qquad j = 1,\ldots,J,$$

$$(3.6)$$

where $\bar{\mathcal{A}}_t = \frac{1}{J}\sum_{j=1}^{J}\mathcal{A}(x_t^{(j)})$. The gradient-based and gradient-free constraint terms coincide in the case where $\mathcal{A}$ is linear.

Remark 3.1: The scheme (3.6) can be rewritten as

$$dx_t^{(j)} = -\frac{1}{J}\sum_{k=1}^{J}\langle \mathcal{G}(x_t^{(k)}) - \bar{\mathcal{G}}_t, \mathcal{G}(x_t^{(j)}) - \tilde{y}\rangle_{\tilde{\Gamma}}(x_t^{(k)} - \bar{x}_t)\,dt, \qquad j = 1,\ldots,J,$$

$$(3.7)$$

where $\mathcal{G}(x) = \big(G(x), x, \mathcal{A}(x)\big)^{\mathsf{T}}$, $\tilde{y} = \big(y, a, 0_d\big)^{\mathsf{T}}$ and

$$\tilde{\Gamma} = \begin{pmatrix} \Gamma & 0 & 0 \\ 0 & \Sigma & 0 \\ 0 & 0 & \nu \end{pmatrix}.$$

Numerically, implementing (3.6) presents the same level of difficulty as implementing unconstrained EKI. The constraint is interpreted here as an additional observation, with very small associated noise.

4. Numerical results for CBO

In this section, we discuss numerical results for CBO with constraints. We first present the numerical scheme employed throughout this section in Sec. 4.1, and then we investigate the qualitative behavior of the method.

In Secs. 4.2 and 4.3, we present numerical results for the case of equality and inequality constraints, respectively. In Sec. 4.4, we investigate the convergence to the mean-field equation (2.8) in the many-particle limit. In Sec. 4.5, we compare the proposed method with another approach in the literature [41,42]. Finally, in Sec. 4.6, we present a brief parameter study, with the aim of highlighting the role of the parameters entering in the method.

In all the numerical experiments presented in this section, the objective function is defined from the Ackley function in two dimension:

$$f_A(x) = -20\exp\left(-\frac{1}{5}\sqrt{\frac{1}{2}\sum_{i=1}^{2}|x_i|^2}\right) - \exp\left(\frac{1}{d}\sum_{i=1}^{2}\cos(2\pi x_i)\right) + e + 20.$$

4.1. *Numerical schemes*

As mentioned, the introduction of a relaxation drift towards the feasible manifold with $\varepsilon \ll 1$ leads to stiffness of the CBO particle system (2.4). This leads to a stringent limitation on the time step when using an explicit method such as Euler–Maruyama [53]. In the case of a single quadratic equality or inequality constraint, this issue can be mitigated by using a semi-implicit scheme. Let us consider, for example, an equality constraint of the form

$$\mathcal{A}(x) = x^{\mathsf{T}} A x - c = 0 \tag{4.1}$$

for a positive definite matrix $A \in \mathbf{R}^{d \times d}$ and a positive real number c. This general form of the constraint encompasses elliptic and, in particular, the circular equality constraints considered throughout this section. Moreover, it is simple to extend the method to similar inequality constraints. In order to use CBO with the constraint (4.1), we propose the discretization

$$y_n^{(j)} = x_n^{(j)} - \left(x_n^{(j)} - m_g(\mu_n^J)\right)\Delta t + \sqrt{2\Delta t}\,\sigma\left(x_n^{(j)} - m_g(\mu_n^J)\right) \circ \xi_n^{(j)}, \tag{4.2a}$$

$$x_{n+1}^{(j)} = \left(I_d + \frac{4\Delta t}{\varepsilon}\mathcal{A}(x_n^{(j)})A\right)^{-1} y_n^{(j)}, \tag{4.2b}$$

where $\xi_n^{(j)} \sim \mathcal{N}(0,1)$, for $n = 1, 2, \ldots$ and $j = 1, \ldots, J$, are independent. This discretization arises when treating the term $\nabla\mathcal{A}$ in (2.4) implicitly and all the other terms on the right-hand side explicitly. The PDE associated with CBO is solved with the help of the Python free software library **Fenics** [54] on a mesh refined locally around the feasible manifold using the open source software **gmsh** [55]. We use the discontinuous Galerkin method with elements of degree zero. The drift term is computed with upwind flux and for the diffusion we use the discretization proposed in Ref. [56].

In order to make the SDE and PDE results comparable, we initialize the particle SDE (2.4)–(2.3) with $x_0^{(j)} \sim \mathcal{N}(0, 3I_2)$, and the corresponding PDE (2.8) is initialized with initial condition $\varrho_0(x) = \frac{1}{\sqrt{2\pi}}e^{-\frac{1}{2}\left(\frac{x}{3}\right)^2}$, i.e. ϱ_0 is the probability density function of $\mathcal{N}(0, 3I_2)$. Unless otherwise stated, we always assume these initial conditions at the particle and PDE levels. The following parameters are fixed for all simulations: $\sigma = 0.7, \alpha = 30, \Delta t_{\mathrm{PDE}} = 5 \times 10^{-3}, \Delta t_{\mathrm{SDE}} = 5 \times 10^{-4}$. The number of particles J, and the relaxation parameter ε, are specified on a case-by-case basis.

4.2. CBO with equality constraint

We now present the results of particle simulations of the CBO method with constraints (2.4), which incorporates both the relaxation drift towards the constraint and the penalization (2.3), and of PDE simulations of the associated mean-field equation (2.8). We begin by considering the following optimization problem with circular equality constraint,

$$\underset{x \in B}{\operatorname{argmin}} f_A(x - x_*), \qquad x_* = \begin{pmatrix} 3 \\ 0 \end{pmatrix}, \qquad B = \{x \in \mathbf{R}^2 : x^2 = 9\}.$$

Here f_A is the Ackley function, which is minimized at $x = 0$, so that the solution of this problem is given by $x_* = (3, 0)$. (Note that this corresponds to the minimizer of the unconstrained global optimization problem.)

The top left panel in Fig. 3 illustrates the shifted Ackley function (filled contour) and the constraint set B (gray line). The top right, bottom left and bottom right panels depict the PDE solution (filled contour), as well as the weighted averages corresponding to 100 independent simulations of the particle system (one red dot per independent simulation), at times 0.01, 0.15 and 0.5. We observe that, already for $t = 0.01$, the weighted averages provide a good approximation of the global minimizer.

4.3. CBO with inequality constraint

Now, we investigate the following example with inequality constraint:

$$\underset{x \in B}{\operatorname{argmin}} f_A(x - x_*), \qquad x_* = \begin{pmatrix} 2 \\ 2 \end{pmatrix}, \qquad B = \{x \in \mathbf{R}^2 : x^2 \geq 18\},$$

where f_A is the standard Ackley function in two dimensions. The associated numerical results are illustrated in Fig. 4. Again, the red particles indicate the positions of the weighted means of $M = 100$ independent simulations with $J = 50$ particles each. The top left panel of Fig. 4 shows the shifted Ackley function (filled contour), and the boundary of the feasible set (gray line). The top right, bottom left and bottom right panels depict the evolution of the PDE solution as well as the weighted means of the SDE simulations. Compared to the previous example, the solution to the mean-field PDE is less concentrated, which is expected since the constraint is of inequality type. At $t = 0.01$ (top right plot), not all weighted averages are close to the minimizer. At $t = 0.15$ and $t = 0.5$ (bottom plots), the weighted averages are very close to the minimizer, and the density concentrates around it.

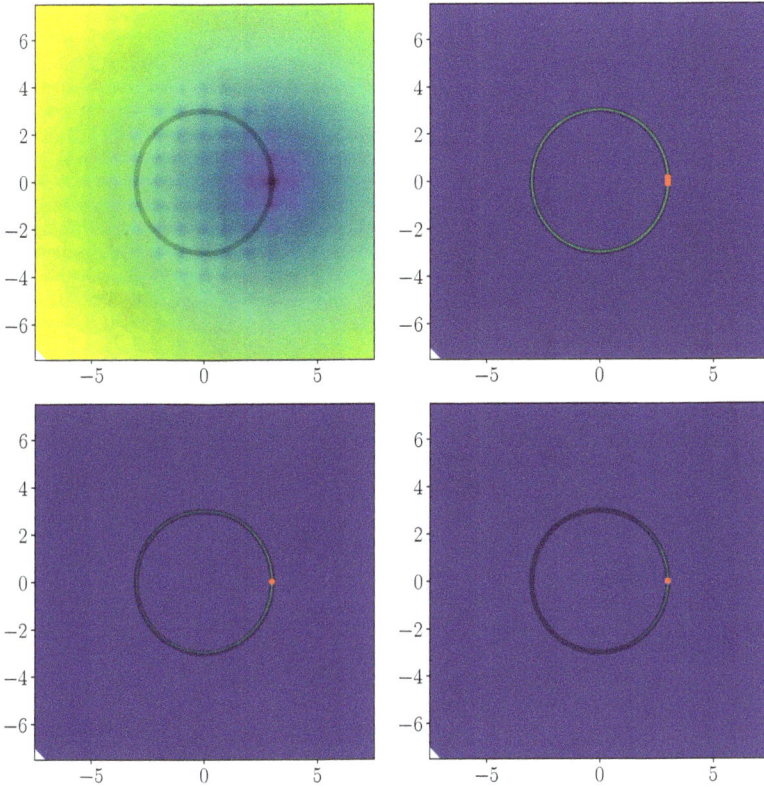

Fig. 3. **Top left panel**: Ackley function shifted by $(3, 0)$, and feasible set $B = \{x \in \mathbf{R}^2 : x^2 = 9\}$. **Other panels**: PDE solution at $t = 0.01$, $t = 0.15$ and $t = 0.5$ (filled contour), and weighted averages of $M = 100$ independent simulations with $J = 50$ particles each (red dots). We observe that the weighted averages lie in the support of the mean-field solution and provide a good approximation of the global minimizer. The parameters employed for the simulation are the following: $J = 50, M = 100, \Delta t_{\mathrm{SDE}} = 0.0005, \Delta t_{\mathrm{PDE}} = 0.005, h_{\min} = 0.125, h_{\max} = 0.5, \alpha = 30, \nu = 1, \sigma = 0.7, \varepsilon = 0.1$.

We emphasize that the accuracy of the PDE results is strongly restricted by the mesh size; since this is fixed during a simulation, the PDE solver is unable to capture the evolution of the ensemble after this has concentrated to a very small area. Comparisons between the SDE and PDE simulations are therefore meaningful only in the beginning of the simulation. For this reason we restrict the following numerical studies to particle simulations.

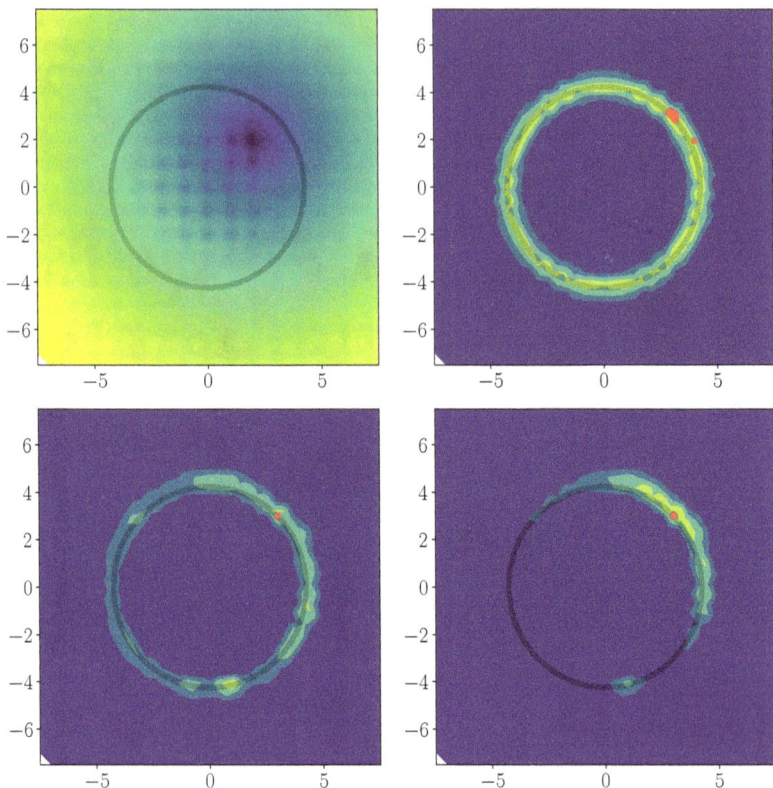

Fig. 4.   Parameters: $J = 50, M = 100, \Delta t_{\mathrm{SDE}} = 0.0005, \Delta t_{\mathrm{PDE}} = 0.005, h_{\min} = 0.125, h_{\max} = 0.5, \alpha = 30, \nu = 1, \sigma = 0.7, \varepsilon = 0.1$ Top-left: Ackley function shifted by $(2, 2)$ and the constraint $x^2 \geq 18$ (gray). Top-right to bottom-right: mean-field density and weighted means of the particle solution at $t = 0.01, 0.15$ and 0.5. The weighted averages lie in the support of the mean-field solution and very close to the global minimum of the function.

4.4. Many particle limit $J \to \infty$ for CBO

In the following we investigate the behavior of the proposed CBO scheme as J tends to infinity. For simplicity, we consider toy examples where the true minimizer x_* is known. For the particle scheme we perform $M = 100$ independent simulations for every setting considered, each with a number J of particles. For $j = 1, \ldots, J$ and $m = 1, \ldots, M$, the vector $x_t^{(m,j)} \subset \mathbf{R}^d$ denotes the position of the j-th particle in the m-th independent simulation at time t. First, we study in Fig. 5 the evolution of the variance (more precisely, the trace of the covariance) of particle positions under an averaged

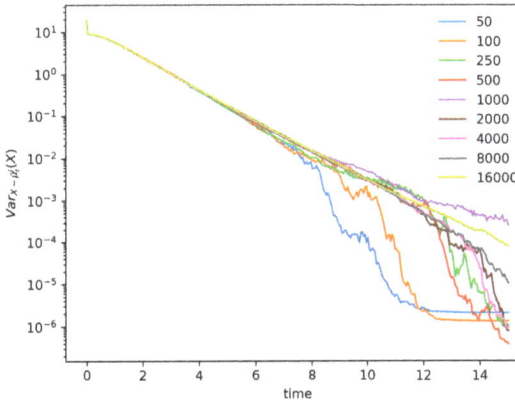

Fig. 5. Ensemble variance for different ensemble sizes. For the simulation we used the following parameters: $M = 100, \Delta t_{SDE} = 0.0005, \alpha = 30, \nu = 1, \sigma = 0.7, \epsilon = 0.1$.

empirical measure,

$$\mathrm{Var}_{X \sim \bar{\mu}_t^J}(X) = \int \left| x - \int y\, \bar{\mu}_t^J(\mathrm{d}y) \right|^2 \bar{\mu}_t^J(\mathrm{d}x), \qquad \bar{\mu}_t^J = \frac{1}{M} \sum_{m=1}^{M} \mu_t^{J,m},$$

where $\mu_t^{J,m}$ denotes the empirical measure for the m-th independent simulation at time t, for different ensemble sizes. During the first few time steps, we observe a fast decrease of all variances, as all the particles quickly move toward the feasible manifold. Thereafter, the ensembles continue concentrating but more slowly. As the figure shows, starting from $t \approx 8$ ensembles with a smaller number of particles begin to collapse faster than ensembles with more particles. This plot does not reveal any information on the proximity of the convergence point to the global minimizer, however.

The accuracy of the convergence point is discussed in Fig. 6, which depicts the time evolution of the expected value of the Wasserstein distance $W_2(\mu_t^J, \delta_{x_*})$, approximated in practice based on 100 independent simulations. Again, we see a strong decrease at the beginning of the simulation as the particles are driven towards the manifold. Then the expected values of the Wasserstein distances decrease exponentially until the ensembles collapse. As we already observed in Fig. 5, smaller ensembles collapse earlier than larger ones. The approximation of the true minimizer is very good already for small ensemble sizes.

To illustrate the convergence to the deterministic mean-field dynamics in the limit as $J \to \infty$, we plot the Wasserstein distances of $W_2(\mu^J, \mu^{16000})$ for different values of J in Fig. 7, with this time only one simulation per

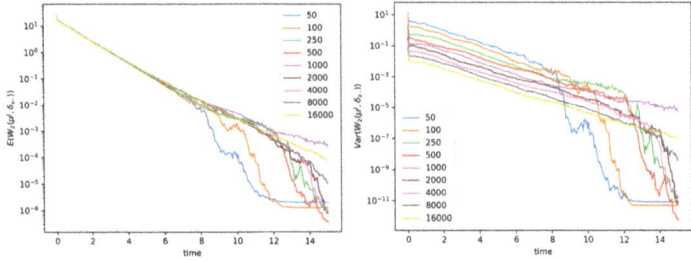

Fig. 6.   Evolution in time of the expectation (**left**) and variance (**right**) of the Wasserstein distances. The parameters employed here are the same as in Fig. 5.

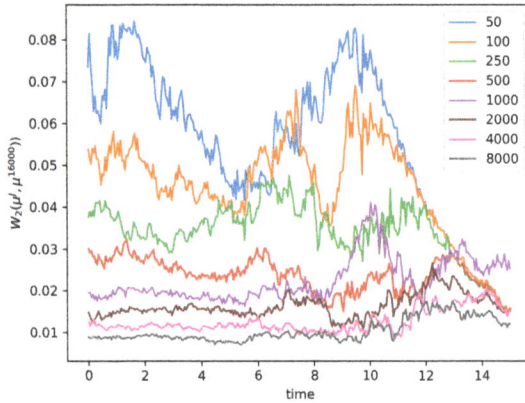

Fig. 7.   Wasserstein distance to reference solution with $J = 16000$. The parameters employed here are the same as in Fig. 5.

value of J. At $t = 0$ we see the expected decrease of the Wasserstein distance along the y-axis. The distances vary over time but decrease at the end of the simulation as all ensembles converge towards the global minimizer.

In Fig. 8 the average distance of the ensembles to the constraint is illustrated with the help of the quantity

$$\mathrm{CE}(t) = \frac{1}{M} \sum_{m=1}^{M} \left(\frac{1}{J} \sum_{i=1}^{J} \mathcal{A}\big(x_t^{(m,j)}\big)^2 \right) \approx \mathbf{E} \left(\int_{\mathbf{R}^d} |\mathcal{A}(x)|^2 \, \mu_t^J(\mathrm{d}x) \right). \quad (4.3)$$

This may be viewed as a "constraint energy"; the smaller $\mathrm{CE}(t)$ is, the closer the ensemble is to the constraint. At the beginning of the simulation, the quantity $\mathrm{CE}(t)$ decreases very fast thanks to the relaxation drift towards the manifold. As time increases, we observe small fluctuations of the constraint energy, but it is likely that these oscillations would disappear in the

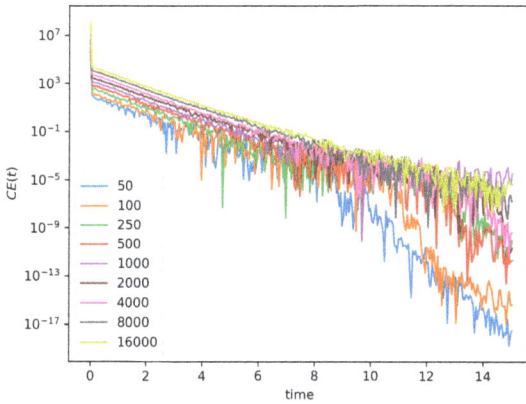

Fig. 8. Distance to constraint measured with the quantity defined in (4.3). The parameters employed here are the same as in Fig. 5.

limit $M \to \infty$ of many independent simulations. The oscillations are due to the diffusion term acting on the particles, which is stronger for particles far away for the weighted mean. At $t = 15$, all ensembles are close to the constraint.

4.5. *Comparison with a projection-based method*

In the introduction, we mentioned that, in the case of CBO, the approach discussed here incorporates the constraint via two distinct mechanisms: a penalization added to the objective function, and a relaxation drift towards the manifold. This is in contrast with other methods that ensure that the constraint is satisfied at all times. We therefore make a comparison of the method proposed here and a method using projection to the constraint in every time step, as proposed in Ref. [41]. The results of the comparison are presented in Fig. 9. In the left panel, which depicts the evolution of the expected value of the variance, we observe that the two graphs differ at the beginning, which is not surprising given that the ensembles are not initially confined to the manifold with our method, and then the two approaches perform comparably. One advantage of our approach is that the initial ensemble can be drawn from any probability distribution over $\mathbf{R}^d$, whereas for the projection method of Ref. [41] the initial ensemble needs to be supported on the constraint. In addition, calculating projections to the feasible manifold at each time step can be computationally costly. The panel in the middle illustrates the evolution of the expected value of the

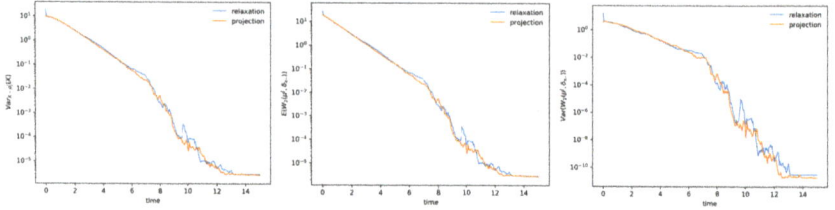

Fig. 9. Comparison of our method (2.4) with the projection-based method. The parameters employed for the simulation are the following: $J = 50, M = 100, \alpha = 30, \nu = 1, \Delta t = 0.0005, \sigma = 0.7, \varepsilon = 0.1$.

Wasserstein distance to the minimizer, $W_2(\mu^{50}, \delta_{x_*})$, approximated for each method from 100 independent simulations and the evolution of the variance of the Wasserstein distance to the minimizer, again approximated for each method from 100 independent simulations, is shown on the right. As expected, the ensembles corresponding to both methods eventually collapse and provide reasonable approximations of the global minimizer.

We notice that, in all the above plots concerning CBO, the time evolution of the particle ensemble can be divided into roughly three phases. In the first phase, at the beginning of the simulation, the particles are quickly driven to the manifold. In the second phase, the particles move along the feasible manifold towards the solution of the constrained optimization problem, and few fluctuations are observed in the plots. The size of the fluctuations, in logarithmic scale, in the final phase are consistent with the fact that the amplitude of the noise decreases as we get closer to the collapse of the ensemble near the global minimizer at the final stages of the simulation.

To conclude this section, we examine the influence of the parameter ε, which enters in the relaxation drift, on the speed of relaxation towards the constraint. This is illustrated in Fig. 10, where the evolution of "constraint energy", measured using (4.3), is shown for different values of ε. For all the values of ε considered, the "constraint energy" is approximated based on $M = 100$ independent simulations with $J = 1000$ particles each.

4.6. Numerical experiments for the CBO particle system

In this section, we present additional numerical experiments concerning the particle system. We investigate, in particular, the influence of the parameters ν, ε and J on the accuracy of the convergence point of CBO, first for

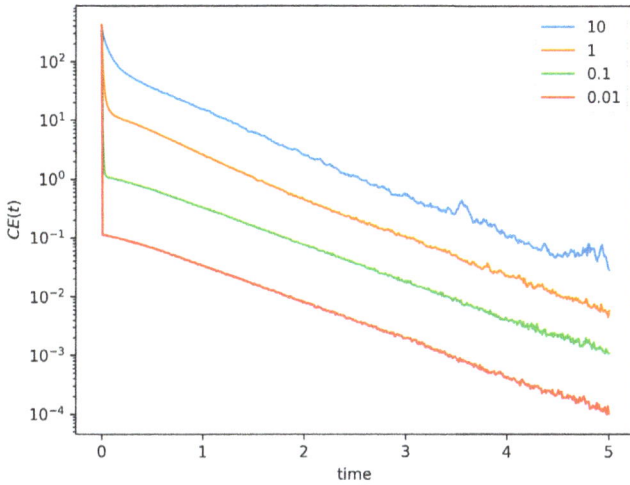

Fig. 10. Distance to constraint for different relaxation parameters ϵ. Other parameters of the simulation are: $J = 1000, M = 100, \alpha = 30, \nu = 1, \Delta t = 0.0005, \sigma = 0.7$.

the following optimization problem:

$$\operatorname*{argmin}_{x \in B} f_A(x - x_*), \quad x_* = \begin{pmatrix} 2 \\ 2 \end{pmatrix}, \quad B = \{x \in \mathbf{R}^2 : x^2 + y^2 = 18\}. \quad (4.4)$$

All the numerical results presented below are generated from CBO with the parameters $\sigma = 0.7$ and $\Delta t = 0.01$. In each simulation, the particles forming the initial ensembles are always drawn from $\mathcal{N}(0, 100I_2)$. The other parameters are specified on a case-by-case basis.

In Fig. 11, we illustrate for different values of $\nu \in \{10, 1, .1, .01\}$ and $J \in \{100, 1000\}$, the convergence points of 100 independent simulations (per set of parameters). Here we take $\varepsilon = \infty$; that is, we employ the CBO particle system with penalization but without relaxation drift. From the figures corresponding to $J = 100$, we observe that, although smaller values of ν enable to enforce the constraint more effectively, taking ν too small is detrimental for the accuracy of the convergence point. This effect is observed also for $J = 1000$, but to a much lesser extent, which may indicate a better behavior of the system in the many-particle limit. This numerical experiment highlights the considerable influence of the number of particles on the behavior of the method. In practice, a trade-off needs to be achieved between precision and computational cost.

In Fig. 12, we illustrate the convergence points of 100 independent simulations with this time $(\nu, \varepsilon) \in \{.1, 10\}^2$, and for fixed $J = 100$.

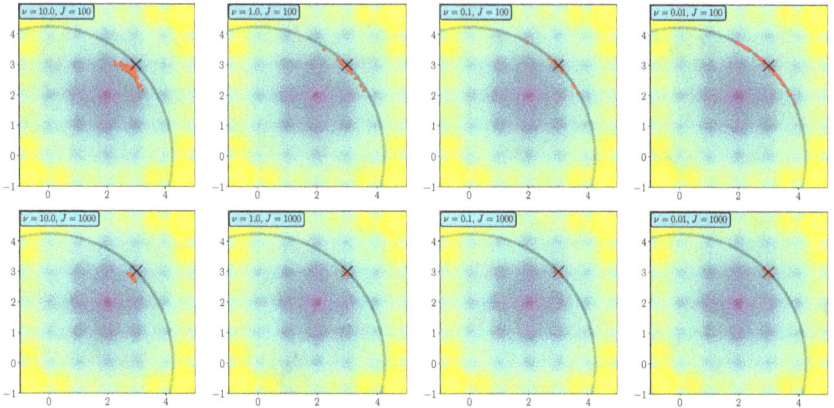

Fig. 11.   Points of convergence of $M = 100$ simulations of CBO for the constrained optimization problem (4.4), for different values of ν and J and for $\varepsilon = 0$. The black cross indicates the position of the global minimizer under the constraint.

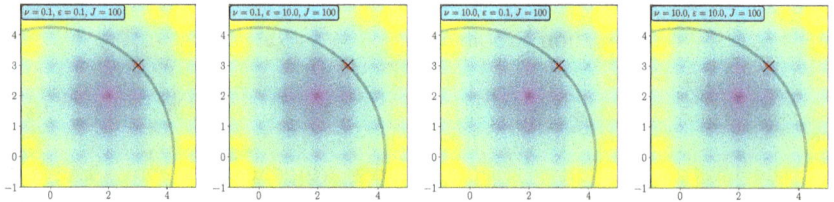

Fig. 12.   Points of convergence of $M = 100$ simulations of CBO for the constrained optimization problem (4.4), for different values of ν and ε.

For the simulations corresponding to $\varepsilon = .1$, we take a smaller time step $\Delta t = 10^{-3}$ in order to avoid stability issues, and in all the simulations, we employ the semi-implicit scheme (4.2). Interestingly, the presence of the relaxation drift, even with a relatively small amplitude $1/\varepsilon$, considerably improves the performance of the method compared to the case without penalization.

To conclude this section, we present in Fig. 13 numerical results for (4.4) with the inequality constraint $B = \{x \in \mathbf{R}^2 \colon x^2 + y^2 \geq 18\}$ instead of an equality constraint. The problem setting is then the same as in Fig. 3. Rather than presenting the points of convergence of the method for several parameter choices as in the previous figures, here we perform only one simulation with a given set of parameters ($\nu = \varepsilon = 1$, $J = 100$), and we illustrate the evolution in time of the ensemble. As the figure shows, the

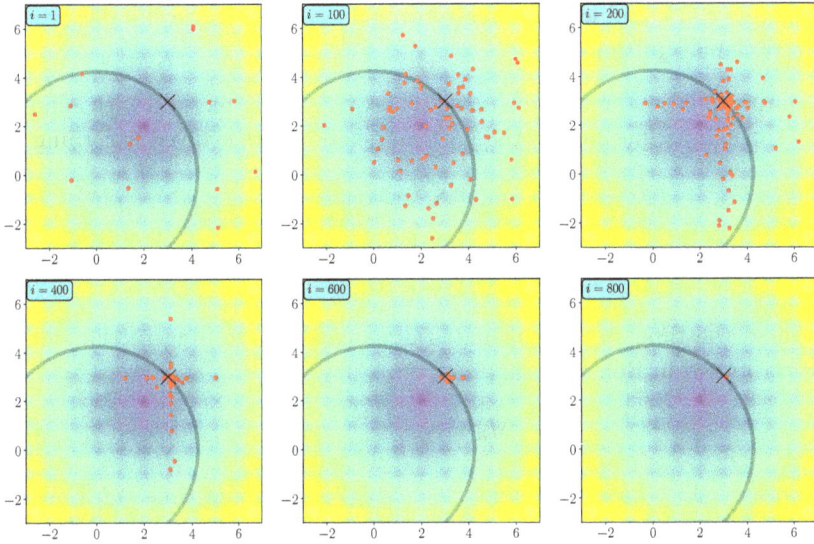

Fig. 13. Evolution of the particles for simulations when CBO is employed for minimizing the Ackley function under the constraint that $\{x^2 + y^2 \geq 18\}$. The parameters for this example are $\varepsilon = \nu = 1$ and $J = 100$.

ensemble appears to collapse after roughly 800 iterations, that is at time $t \approx 8$. In addition, the point of convergence is very close to the optimal solution given the constraint.

5. Numerical results for EKI with constraints

In this section, we present numerical results for EKI with constraints. We simulate only the particle system and do not present result for the associated mean-field equation. Our numerical experiments aim at illustrating the performance of the method on a toy problem, as a proof of concept.

The numerical schemes employed are described in Sec. 5.1, and numerical results for a simple inverse problem arising from a PDE application are presented in Sec. 5.2.

5.1. *Numerical schemes*

The presence of the relaxation term originating from the constraint poses a challenge also for the discretization of EKI with constraints (3.5), although to a lesser extent than for CBO thanks to the affine-invariance of the method [34]. In this section, we consider specifically the derivative-free dynamics (3.7), and we investigate two different approaches for discretizing this dynamics in time:

- An explicit discretization using the explicit Euler method with adaptive time step. Specifically, the particle positions are evolved according to

$$x_{n+1}^{(j)} = x_n^{(j)} - \Delta t_n \sum_{k=1}^{J} M_n^{kj} x_n^{(k)}, \qquad j = 1, \ldots, J,$$

$$M_n^{kj} = \frac{1}{J} \langle \mathcal{G}(x_n^{(k)}) - \bar{\mathcal{G}}_n, \mathcal{G}(x_n^{(j)}) - \tilde{y} \rangle_{\tilde{\Gamma}}, \qquad (5.1)$$

and the time step is calculated dynamically according to the method proposed in Ref. [36], that is

$$\Delta t_n = \frac{\Delta t_*}{\|M_n\|_2 + \frac{\Delta t_*}{\Delta t_{\max}}}, \qquad (5.2)$$

where Δt_* (base time step) and $\Delta t_{\max}$ (maximum time step) are parameters, and where M_n is the matrix with entries M_n^{kj}. With the notation $X_n = (x_n^{(1)}, \ldots, x_n^{(J)})$, Eq. (5.1) reads in matrix form as

$$X_{n+1} = X_n - \Delta t_n X_n M_n. \qquad (5.3)$$

- A semi-implicit discretization given by

$$X_{n+1} = X_n - \Delta t_n X_{n+1} M_n,$$

where the same notation is used as in the previous item. The associated update formula is given by

$$X_{n+1} = X_n (I_d + \Delta t_n M_n)^{-1}. \qquad (5.4)$$

Although this approach could be employed with a fixed time step, we obtain faster convergence when the time step is adapted according to (5.2), and so we consider only the latter setting.

Remark 5.1: Note that $\sum_{k=1}^{K} M_n^{kj} = 0$ for all n and j, by definition of $\bar{\mathcal{G}}_n$. Consequently, the update formulas (5.3) and (5.4) can be rewritten equivalently as

$$X_{n+1} = X_n - \Delta t_n (X_n - \bar{x}_n) M_n, \qquad (5.5a)$$
$$X_{n+1} = \bar{x}_n + (X_n - \bar{x}_n)(I_d + \Delta t_n M_n)^{-1}, \qquad (5.5b)$$

where $\bar{x}_n = \frac{1}{J} \sum_{j=1}^{J} x_n^{(j)}$ and $X_n - \bar{x}_n$ denote, by a slight abuse of notation, the matrix obtained by subtracting $\bar{x}_n$ from each column of X_n. Although mathematically equivalent, these discretizations are observed in our numerical experiments to be much less sensitive to roundoff errors, and therefore preferable in practice.

5.2. *Numerical experiments for EKI with constraints*

In this section, we illustrate the performance of the dynamics (3.7), discretized according to (5.5a) or (5.5b), for solving a toy inverse problem with constraints. The problem considered concerns the recovery of the initial condition of a Fokker–Planck equation based on incomplete observation of the solution. More precisely, our aim is to find the parameters of a Gaussian mixture

$$\varrho_0(x) = \sum_{n=1}^{N} w_n \frac{\exp\left(-\dfrac{|x - m_n|^2}{2\sigma_n^2}\right)}{\sqrt{2\pi\sigma_n^2}}, \qquad (5.6)$$

given noisy observations at time T of the solution $\varrho(x, t)$ to the following initial value problem with initial condition ϱ_0:

$$\begin{cases} \partial_t \varrho = \partial_x(x\varrho + \partial_x\varrho), & \text{in } \mathbf{R} \times (0, T), \\ \varrho = \varrho_0, & \text{on } \mathbf{R} \times \{t = 0\}. \end{cases} \qquad (5.7)$$

We assume that the data is composed of noisy observations

$$y_k = \varrho(x_k, T) + \eta_k, \qquad 1 \le i \le K,$$

where $x_k \in \mathbf{R}$ are discrete positions and η_k are noise terms drawn independently from $\mathcal{N}(0, \gamma^2)$ for some covariance γ^2. The observation positions are given by $x_k = -L + (k-1)\frac{L}{K-1}$, i.e. they are uniformly spread between $-L$ and L, with both ends included.

Recovering the weights. Assuming first that the means and variances of the Gaussian mixture (5.6) are known, we seek to find, as an approximation of the true weights $w^\dagger$, a vector of weights $w = (w_1, \ldots, w_N)^\mathsf{T}$ that minimizes the least-square misfit

$$f(w) = \sum_{k=1}^{K} \frac{1}{\gamma^2} |y_k - \varrho_e(x_k, T; w)|^2,$$

under the constraints that $\sum_{n=1}^{N} w_n = 1$ and $w_n \ge 0$ for all $n \in \{1, \ldots N\}$, which guarantee that ϱ_0 is a probability density. Here $\varrho_e(T, x; w)$ is the exact solution to the initial value problem (5.7) when the weights in the initial condition are given by w. This admits the following explicit expression:

$$\varrho_e(x, t; w) = \sum_{n=1}^{N} w_n \frac{\exp\left(-\dfrac{|x - m_n(t)|^2}{2\sigma_n(t)^2}\right)}{\sqrt{2\pi\sigma_n(t)^2}}, \qquad (5.8)$$

with $m_n(t) = e^{-t}m_n$ and $\sigma_n(t) = \sqrt{1+(\sigma_n^2-1)e^{-2t}}$. Note that, in this problem, the forward model $w \mapsto \{\varrho_e(x_k,T;w)\}_{k=1}^K$ is linear and so, in the absence of constraints, the ensemble Kalman inversion method performs an exact preconditioned gradient descent.

We assume, for simplicity, that the Gaussian mixture is composed of only $N = 2$ components, and that the known means and variances of these components are given by $m_1 = -m_2 = 4$ and $\sigma_1 = \sigma_2 = .1$. For the observations, we take the parameters $L = 10$, $\gamma = .01$ and $K = 100$, and for the adaptation of the time step (5.2) we use the parameters $\Delta t_* = 1$ and $\Delta t_{max} = \infty$. The ensemble size is taken to be $J = 100$, the final time is $T = .5$, and the noisy observations are illustrated in Fig. 14.

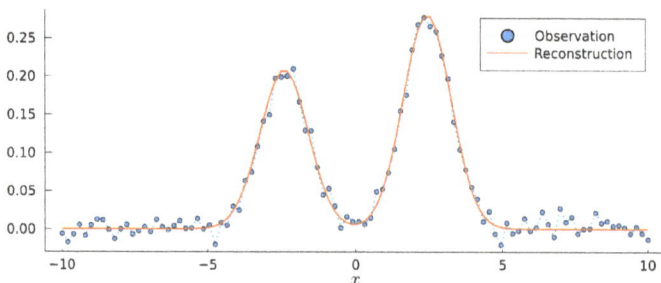

Fig. 14. Noisy observations of the solution to the Fokker–Planck equation, in the case of only 2 components in the binary mixture, and solution at time T obtained by using the reconstructed initial condition.

We begin by analyzing the influence of the small parameter ν on the solution returned by the method when the particles forming the initial ensemble, each corresponding to the couple of weights, are drawn from $\mathcal{N}(0,I_2)$. For this numerical experiment, we use only the explicit discretization (5.5a). Since the evolution (3.7) is deterministic, and since we observed empirically that the initial ensemble does not have much influence on the point of convergence of the method, we run only one simulation per value of ν. In the table below, we indicate for several values of ν the value of the sum $w_1 + w_2$ at the point of convergence of the method, as well as the distance to the true value of the weights from which the observed data was generated. This was taken to be $w^\dagger = (0.411..., 0.588...)$.

ν	$w_1 + w_2$	$|w_1 - w_1^\dagger| + |w_2 - w_2^\dagger|$
1	0.9845...	0.0282...
10^{-2}	0.9847...	0.0282...
10^{-4}	0.99289...	0.0282...
10^{-6}	0.999869...	0.0282...
10^{-8}	0.99999868...	0.0282...

As expected, we observe that the smaller ν, the closer the point of convergence is to the feasible manifold. In addition, the proximity to the true weights is hardly affected by changes in the value of ν. Figures 15 and 16 depict the evolution of the ensemble for $\nu = 1$ and $\nu = 10^{-8}$. Although the method converges quickly in both cases, the dynamics look qualitatively very different: for $\nu = 1$ the effect of the penalization is not clearly apparent, while for $\nu = 10^{-8}$ the constraint term appears dominant in the penalized objective function (2.3) depicted in the background. In the latter case, the particles first converge to a vicinity of the feasible line $w_1 + w_2 = 1$, and then move along this line to the optimizer.

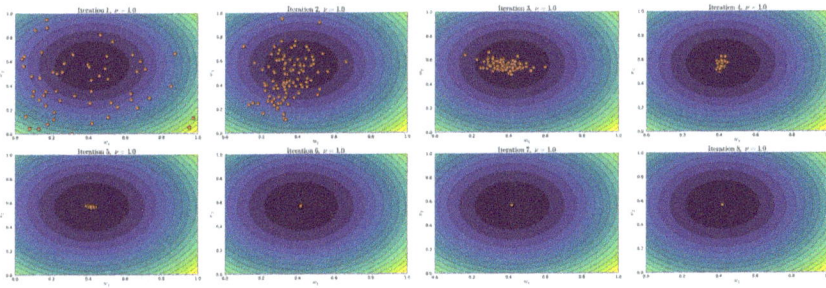

Fig. 15. Evolution of the ensemble obtained from the explicit discretization (5.5a) of EKI with constraints, in the case where $\nu = 1$. The penalized objective function (denoted by g in (2.3)) is depicted as a filled contour in the background.

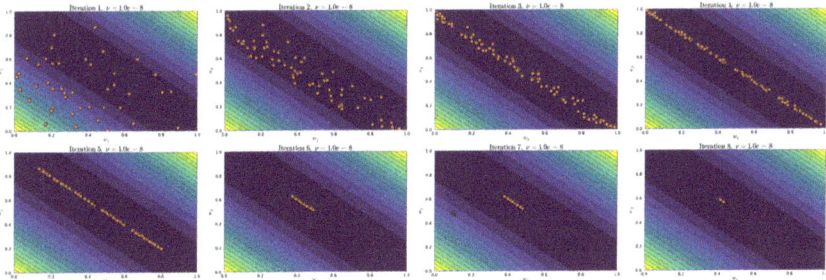

Fig. 16. Like Fig. 15, this plot depicts the evolution the discrete-time dynamics (5.5a), with now $\nu = 10^{-8}$.

To conclude this paragraph, we compare the two discretizations proposed in (5.1), when both methods use the same initial ensemble and with $\nu = 10^{-8}$. In the left panel of Fig. 17, we illustrate the evolution of the error, measured as $|\bar{w}_1 - w_1^\dagger| + |\bar{w}_2 - w_2^\dagger|$, where $\bar{w}_1$ are $\bar{w}_2$ are sample averages of the weights over the particles. In the right panel, we present the evolution of the matrix 2-norm of the sample covariance for each method. Both methods converge quickly, but the figures indicate a slightly faster convergence for the fully explicit discretization.

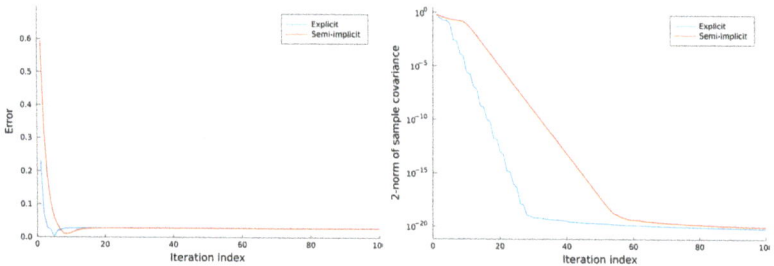

Fig. 17. Evolution of the error (**left**) and sample covariance (**right**), for the explicit and semi-implicit discretizations described in Sec. 5.1. In both cases, the time step is adapted dynamically according to (5.2). We emphasize that the error is computed with respect not to the optimal solution to (5.9) but to the true weights, which were employed to generate the data. It is not surprising, therefore, that the error converges to a strictly positive value in the limit as the iteration index tends to infinity.

Recovering the weights and variances. In this paragraph, we consider the more challenging case where $N = 3$ and both the weights and variances of the initial Gaussian mixture need to be recovered, in which case the forward model is no longer linear. For simplicity, we assume that the means $\{m_n\}_{n=1}^N$ are still known and this time given by $(m_1, m_2, m_3) = (-5, 0, 5)$, and we seek to minimize

$$f(w, v) = \sum_{k=1}^K \frac{1}{\gamma^2} \left| y_k - \varrho_e \left(x_k, T; w, \mathrm{abs.}(v) \right) \right|^2, \qquad (5.9)$$

where w is the vector of weights and $v = (\sigma_1^2, \sigma_2^2, \sigma_3^2)$ is the vector of variances, under the constraints

$$\sum_{n=1}^3 w_n = 1, \qquad w_n \geq 0 \quad \forall n \in \{1, 2, 3\}, \qquad \sigma_n^2 \geq 0 \quad \forall n \in \{1, 2, 3\}.$$

Here the function "abs." denotes the element-wise absolute value, and the function $(x, t) \mapsto \varrho_e(x, t; w, v)$ denotes the exact solution (5.8) to the Fokker–Planck equation (5.7) when the parameters w and v are employed in the initial condition (5.6). We define the objective function in this manner, with the presence of the "abs." function, in order to guarantee that this function can be evaluated for any choice of parameters $w \in \mathbf{R}^3$ and $v \in \mathbf{R}^3$, which is a requirement for applying EKI.

Apart from the parameters of the Gaussian mixtures, and unless otherwise specified, all the parameters employed to generate the numerical results presented in the rest of this section are the same as in the previous paragraph. We begin by examining the influence of the small parameter ν on the error and convergence point of the method. We employ to this end the explicit method described in Sec. 5.1. The table below, in which all figures are truncated after three significant digits, gives the points of convergence of EKI with constraint for different values of ν. We observe again that small values of ν lead to a point of convergence closer to the feasible manifold, as is expected.

	$w_1 + w_2 + w_3$	w_1	w_2	w_3	σ_1^2	σ_2^2	σ_3^2
Truth	1	0.333	0.476	0.191	0.400	0.100	0.500
$\nu = 1$	0.9844	0.337	0.480	0.167	0.433	0.095	0.303
$\nu = 10^{-2}$	0.9847	0.337	0.480	0.167	0.434	0.095	0.305
$\nu = 10^{-4}$	0.99502	0.342	0.482	0.171	0.467	0.093	0.363
$\nu = 10^{-6}$	0.9999273	0.344	0.483	0.173	0.483	0.092	0.392
$\nu = 10^{-8}$	0.999999270	0.344	0.483	0.173	0.483	0.092	0.393

The solution to the Fokker–Planck equation at time T, when using as parameters in the initial condition the point of convergence of EKI with small parameter $\nu = 10^{-8}$, is illustrated in Fig. 18. A good qualitative agreement between the observed and reconstructed solutions is observed.

To conclude this section, we compare in Fig. 19 the two discretization methods (5.5a) and (5.5b) in the case where $\nu = 10^{-8}$. All the ensemble members are initialized according to $\mathcal{N}(0, I_6)$. The error depicted in the left panel is computed as

$$\sum_{n=1}^{3} |\bar{w}_n - w_n^\dagger| + |\bar{\sigma}_n^2 - (\sigma_n^2)^\dagger|,$$

where $w_n^\dagger$ and $(\sigma_n^2)^\dagger$ are respectively the true weights and variances of the components of the initial Gaussian mixture, whereas $\bar{w}_n$ and $\bar{\sigma}_n^2$ are the average weights and variances over the ensemble. As already remarked in

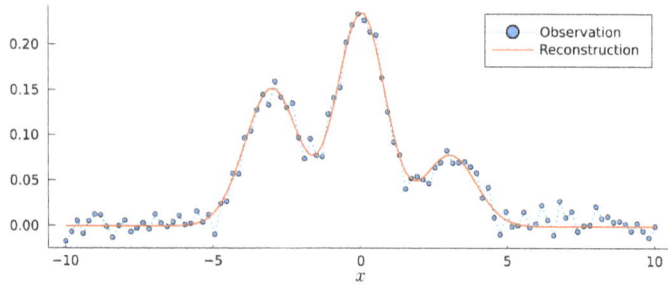

Fig. 18.   Noisy observations (blue dots) and solution to the Fokker–Planck equation (5.7) at time T when using the reconstructed initial condition.

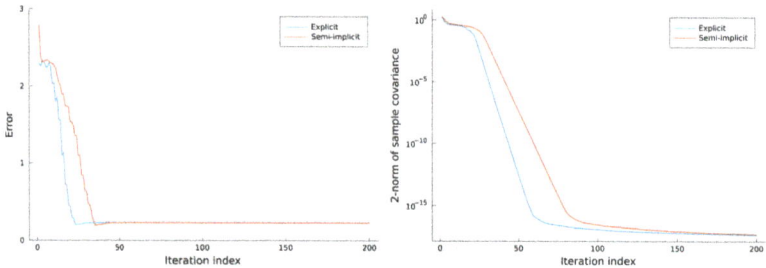

Fig. 19.   Evolution of the error (**left**) and sample covariance (**right**), for the explicit and semi-implicit discretizations described in Sec. 5.1. In both cases, the time step is adapted dynamically according to (5.2).

the caption of Fig. 17, the error does not decrease to 0 as the number of iterations is increased, because the true value of the mixture parameters do not coincide with the minimizer of $f(w, v)$ in (5.9). As shown in the right panel, the sample variances decrease to a small value below 10^{-15} in less than a hundred iterations.

6. Conclusions

In this chapter, we study a new approach for incorporating constraints in consensus-based optimization and ensemble Kalman methods. We demonstrate that, despite their simplicity and ease of implementation, the proposed methods perform well for a number of toy examples. Our study is mostly qualitative and aims at providing a general idea of the performance of the methods, their behavior in the many-particle limit (in the case of CBO), and the influence of the parameters they contain.

Future investigation would be useful in order to determine the practical value of the proposed approach for realistic high-dimensional constrained optimization problems, and to further assess its efficiency in comparison with other methods in the literature. It would also be worthwhile to obtain quantitative results on the link between the particle systems and their mean-field limits, which could inform the choice of the number of particles employed in practice.

Acknowledgments

JAC was supported by the Advanced Grant Nonlocal-CPD (Nonlocal PDEs for Complex Particle Dynamics: Phase Transitions, Patterns and Synchronization) of the European Research Council Executive Agency (ERC) under the European Union's Horizon 2020 research and innovation programme (grant agreement No. 883363) and by EPSRC grant number EP/T022132/1. JAC and UV were also supported by EPSRC grant number EP/P031587/1. CT was partly supported by the European social Fund and by the Ministry Of Science, Research and the Arts Baden-Württemberg (Germany). UV was also supported by the Fondation Sciences Mathématiques de Paris (FSMP), through a postdoctoral fellowship in the "mathematical interactions" program.

References

1. E. Aarts and J. Korst, *Simulated annealing and Boltzmann machines*. New York, NY; John Wiley and Sons Inc., 1988.
2. T. Back, D. B. Fogel and Z. Michalewicz, *Handbook of evolutionary computation*. IOP Publishing Ltd., 1997.
3. C. Reeves, *Genetic algorithms*. Springer, 2003.
4. C. Blum and A. Roli, Metaheuristics in combinatorial optimization: Overview and conceptual comparison, *ACM Computing Surveys (CSUR)*. **35** (2003), 268–308.
5. M. Dorigo and C. Blum, Ant colony optimization theory: A survey, *Theoretical computer science*. **344** (2005), 243–278.
6. J. Kennedy, Particle swarm optimization. In *Encyclopedia of Machine Learning*, pp. 760–766. Springer, 2010.
7. B. Chandra Mohan and R. Baskaran, A survey: Ant colony optimization based recent research and implementation on several engineering domain, *Expert Systems with Applications*. **39** (2012), 4618–4627.
8. D. Karaboga, B. Gorkemli, C. Ozturk and N. Karaboga, A comprehensive survey: artificial bee colony (abc) algorithm and applications, *Artificial Intelligence Review*. **42** (2014), 21–57.

9. X.-S. Yang. Firefly algorithms for multimodal optimization. In eds. O. Watanabe and T. Zeugmann, *Stochastic Algorithms: Foundations and Applications*, pp. 169–178, Springer Berlin Heidelberg, Berlin, Heidelberg, 2009.

10. Z. Bayraktar, M. Komurcu, J. A. Bossard and D. H. Werner, The wind driven optimization technique and its application in electromagnetics, *IEEE transactions on antennas and propagation.* **61** (2013), 2745–2757.

11. J. Kennedy and R. Eberhart. Particle swarm optimization. In *Proceedings of ICNN'95-international conference on neural networks*, vol. 4, pp. 1942–1948, IEEE, 1995.

12. R. Poli, J. Kennedy and T. Blackwell, Particle swarm optimization, *Swarm intelligence.* **1** (2007), 33–57.

13. L. Bianchi, M. Dorigo, L. M. Gambardella and W. J. Gutjahr, A survey on metaheuristics for stochastic combinatorial optimization, *Natural Computing: an international journal.* **8** (2009), 239–287.

14. R. Holley and D. Stroock, Simulated annealing via Sobolev inequalities, *Communications in Mathematical Physics.* **115** (1988), 553–569.

15. R. A. Holley, S. Kusuoka and D. W. Stroock, Asymptotics of the spectral gap with applications to the theory of simulated annealing, *Journal of functional analysis.* **83** (1989), 333–347.

16. C.-R. Hwang and S.-J. Sheu, Large-time behavior of perturbed diffusion markov processes with applications to the second eigenvalue problem for Fokker-Planck operators and simulated annealing, *Acta Applicandae Mathematica.* **19** (1990), 253–295.

17. D. Márquez, Convergence rates for annealing diffusion processes, *The Annals of Applied Probability.* (1997), 1118–1139.

18. M. Pelletier, Weak convergence rates for stochastic approximation with application to multiple targets and simulated annealing, *Annals of Applied Probability.* (1998), 10–44.

19. H. Robbins and S. Monro, A stochastic approximation method, *The annals of mathematical statistics.* (1951), 400–407.

20. L. Bottou et al., Online learning and stochastic approximations, *On-line learning in neural networks.* **17** (1998), 142.

21. S. Bubeck, Convex optimization: Algorithms and complexity, *Foundations and Trends® in Machine Learning.* **8** (2015), 231–357.

22. S. Jin, L. Li and J.-G. Liu, Random batch methods (RBM) for interacting particle systems, *J. Comput. Phys.* **400** (2020), 108877, 30.

23. S. Jin, L. Li and J.-G. Liu, Convergence of the random batch method for interacting particles with disparate species and weights, *SIAM J. Numer. Anal.* **59** (2021), 746–768.

24. R. Pinnau, C. Totzeck, O. Tse and S. Martin, A consensus-based model for global optimization and its mean-field limit, *Math. Models Methods Appl. Sci.* **27** (2017), 183–204.

25. J. A. Carrillo, Y.-P. Choi, C. Totzeck and O. Tse, An analytical framework for consensus-based global optimization method, *Mathematical Models and Methods in Applied Sciences.* **28** (2018), 1037–1066.

26. J. A. Carrillo, S. Jin, L. Li and Y. Zhu, A consensus-based global optimization

method for high dimensional machine learning problems, *ESAIM. Control, Optimisation and Calculus of Variations.* **27** (2021), Paper No. S5, 22.

27. J. Carrillo, F. Hoffmann, A. Stuart and U. Vaes, Consensus based sampling, *arXiv e-prints.* **2106.02519** (2021).

28. C. Totzeck, Trends in consensus-based optimization, *arXiv e-prints.* **2104.01383** (2021).

29. Y. Chen and D. S. Oliver, Ensemble randomized maximum likelihood method as an iterative ensemble smoother, *Mathematical Geosciences.* **44** (2012), 1–26.

30. A. A. Emerick and A. C. Reynolds, Investigation of the sampling performance of ensemble-based methods with a simple reservoir model, *Computational Geosciences.* **17** (2013), 325–350.

31. M. A. Iglesias, K. J. H. Law and A. M. Stuart, Ensemble Kalman methods for inverse problems, *Inverse Problems.* **29** (2013), 045001, 20.

32. C. Schillings and A. M. Stuart, Analysis of the ensemble Kalman filter for inverse problems, *SIAM J. Numer. Anal.* **55** (2017), 1264–1290.

33. A. Garbuno-Inigo, F. Hoffmann, W. Li and A. M. Stuart, Interacting Langevin diffusions: gradient structure and ensemble Kalman sampler, *SIAM J. Appl. Dyn. Syst.* **19** (2020), 412–441.

34. A. Garbuno-Inigo, N. Nüsken and S. Reich, Affine invariant interacting Langevin dynamics for Bayesian inference, *SIAM Journal on Applied Dynamical Systems.* **19** (2020), 1633–1658.

35. J. A. Carrillo and U. Vaes, Wasserstein stability estimates for covariance-preconditioned Fokker-Planck equations, *Nonlinearity.* **34** (2021), 2275–2295.

36. N. B. Kovachki and A. M. Stuart, Ensemble Kalman inversion: a derivative-free technique for machine learning tasks, *Inverse Problems.* **35** (2019), 095005, 35.

37. A. B. Duncan, A. M. Stuart and M.-T. Wolfram, Ensemble Inference Methods for Models With Noisy and Expensive Likelihoods, *arXiv e-prints.* **2104.03384** (2021).

38. G. A. Pavliotis, *Stochastic processes and applications.* vol. 60, *Texts in Applied Mathematics*, Springer, New York, 2014. ISBN 978-1-4939-1322-0; 978-1-4939-1323-7.

39. P. D. Miller, *Applied asymptotic analysis.* vol. 75, American Mathematical Soc., 2006.

40. A. Dembo and O. Zeitouni, *Large deviations techniques and applications.* vol. 38, Springer Science & Business Media, 2009.

41. M. Fornasier, H. Huang, L. Pareschi and P. Sünnen, Consensus-based optimization on hypersurfaces: well-posedness and mean-field limit, *Mathematical Models and Methods in Applied Sciences.* **30** (2020), 2725–2751.

42. M. Fornasier, H. Huang, L. Pareschi and P. Sünnen, Consensus-based Optimization on the Sphere II: Convergence to Global Minimizers and Machine Learning, *arXiv e-prints.* **2001.11988** (2020).

43. M. Fornasier, T. Klock and K. Riedl, Consensus-based optimization methods converge globally in mean-field law, *arXiv e-prints.* **2103.15130** (2021).

44. H. Zhang, S. J. Reddi and S. Sra, Riemannian svrg: Fast stochastic optimization on Riemannian manifolds, *Advances in Neural Information Processing Systems.* **29** (2016), 4592–4600.

45. M. Weber and S. Sra. Projection-free nonconvex stochastic optimization on Riemannian manifolds, 2021.

46. D. J. Albers, P.-A. Blancquart, M. E. Levine, E. Esmaeilzadeh Seylabi and A. Stuart, Ensemble Kalman methods with constraints, *Inverse Problems.* **35** (2019), 095007, 28.

47. N. K. Chada, C. Schillings and S. Weissmann, On the Incorporation of Box-Constraints for Ensemble Kalman Inversion, *arXiv e-prints.* **1908.00696** (2019).

48. M. Herty and G. Visconti, Continuous limits for constrained ensemble Kalman filter, *Inverse Problems.* **36** (2020), 075006, 28.

49. M. Bostan and J. A. Carrillo, Asymptotic fixed-speed reduced dynamics for kinetic equations in swarming, *Mathematical Models and Methods in Applied Sciences.* **23** (2013), 2353–2393.

50. M. Jamil and X.-S. Yang, A literature survey of benchmark functions for global optimisation problems, *International Journal of Mathematical Modelling and Numerical Optimisation.* **4** (2013), 150–194.

51. H. Huang and J. Qiu, On the mean-field limit for the consensus-based optimization, *arXiv e-prints.* **2105.12919** (2021).

52. A. M. Stuart, Inverse problems: a Bayesian perspective, *Acta Numer.* **19** (2010), 451–559.

53. D. Higham, An algorithmic introduction to numerical simulation of stochastic differential equations, *SIAM Review.* **43**, 525–546.

54. M. Alnæs, J. Blechta, J. Hake, A. Johansson, B. Kehlet, A. Logg, C. Richardson, J. Ring, M. E. Rognes and G. N. Wells, The FEniCS project version 1.5, *Archive of Numerical Software.* **3** (2015).

55. C. Geuzaine and J.-F. Remacle, Gmsh: a three-dimensional finite element mesh generator with built-in pre- and post-processing facilities, *International Journal for Numerical Methods in Engineering.* **79** (2009), 1309–1331.

56. D. V. Kulkarni, D. V. Rovas and D. A. Tortorelli. A discontinuous Galerkin formulation for solution of parabolic equations on nonconforming meshes. In *Domain decomposition methods in science and engineering XVI*, vol. 55, *Lect. Notes Comput. Sci. Eng.*, pp. 651–658. Springer, Berlin, 2007.